COGNITIVE CODE

COGNITIVE CODE

POST-ANTHROPOCENTRIC INTELLIGENCE
AND THE INFRASTRUCTURAL BRAIN

JOHANNES BRUDER

McGill-Queen's University Press
Montreal & Kingston • London • Chicago

ISBN 978-0-7735-5916-5 (cloth)
ISBN 978-0-7735-5917-2 (paper)
ISBN 978-0-7735-5969-1 (ePDF)
ISBN 978-0-7735-5970-7 (ePUB)

Legal deposit fourth quarter 2019
Bibliothèque nationale du Québec

Printed in Canada on acid-free paper that is 100% ancient forest free
(100% post-consumer recycled), processed chlorine free

This book has been published with the help of a grant from the Institute of
Experimental Design and Media Cultures, University of Applied Sciences and Arts
Northwestern Switzerland.

Funded by the Financé par le
Government gouvernement
of Canada du Canada

Canada Council Conseil des arts
for the Arts du Canada

We acknowledge the support of the Canada Council for the Arts.
Nous remercions le Conseil des arts du Canada de son soutien.

Library and Archives Canada Cataloguing in Publication

Title: Cognitive code : post-anthropocentric intelligence and the
infrastructural brain / Johannes Bruder.
Names: Bruder, Johannes, 1983– author.
Description: Includes bibliographical references and index.
Identifiers: Canadiana (print) 20190198206 | Canadiana (ebook) 20190198257
  | ISBN 9780773559165 (cloth) | ISBN 9780773559172 (paper) | ISBN
  9780773559691 (ePDF) | ISBN 9780773559707 (ePUB)
Subjects: LCSH: Neurosciences. | LCSH: Artificial intelligence. |
  LCSH: Cognitive neuroscience. | LCSH: Computational neuroscience.
Classification: LCC QP355.2.B78 2019 | DDC 612.8/2330285—dc23

This book was typeset in 10.5/13 Sabon.

Contents

Figures

Acknowledgments

I think of this book as a product of various, intersecting institutional lives that have contributed to the interdisciplinary approach of this book. The research that undergirds my discussion of brain imaging epistemologies began during my time at the National Competence Center for Research on Iconic Criticism (or *eikones*) at the University of Basel, where I had many illuminating discussions and exchanges about imaging and modelling in various fields. I thank Inge Hinterwaldner, Reinhard Wendler, David Magnus, Iris Laner, Il-Tschung Lim, Sophie Schweinfurth, and Leon Wansleben for their generosity. Without the support of my former office mate Thomas Brandstetter, I would probably never have started to write this book. He will always occupy a special place deep down in my amygdala. Of all the people I cooperated with during that time, Svenja Matusall has had the most profound impact on this book. Our discussions and the workshops we organized together were rewarding and reassuring at the same time.

A number of researchers who were part of or frequented the Department of Social Science, Health and Medicine at Kings College London were of utmost importance in regard to the sociological and anthropological backbone of this book, most importantly Des Fitzgerald, Nicole Batsch, Morten Hillgard Bülow, Angela Filipe, Cathy Herbrand, Claire Marris, Erika Mansnerus, Tara Mahfoud, and Luna Rodrigues Silva. Special thanks go to the members of the European Neuroscience and Society Network and to the attendees of its Neuroschool in Bergen.

A few years ago, I took up residence in the Critical Media Lab of the Institute of Experimental Design and Media Cultures in the

Academy of Art and Design at the University of Applied Sciences and Arts Northwestern Switzerland in Basel. I am indebted to the members of "the Lab" for all the discussions we had and for all the welcome distraction they provided. To name but a few, they include Jamie Allen, Leonie Häsler, Felix Gerloff, Moritz Greiner-Petter, Shintaro Miyazaki, Samuel Hanselmann, Flavia Caviezel, Nuria Barcelo, Jan Torpus, and Christiane Heibach. Above all, I thank Claudia Mareis for her generous support and for opening up both new intellectual and new professional perspectives.

I am also grateful to a cast of researchers who were instrumental in the process of finishing this book. Chris Salter, Maya Indira Ganesh, Clemens Apprich, Luciana Parisi, Johanna Schindler, and the members of the Institute of Advanced Studies "Media Cultures of Computer Simulation," Leuphana University helped me to sharpen my arguments. The discussions on "smartness" that I had with Orit Halpern over the past three years are of particular significance; she was the first to read the entire manuscript and provided me with an unanticipated reading of what I had written that differed from my own. I also thank the anonymous reviewers of the manuscript, who had a similar influence on the completion of the book. Their readings opened new avenues, and Khadija Coxon and the team at McGill-Queen's University Press helped me to cut to the chase.

I am, of course, particularly grateful to everyone who agreed to be interviewed or "observed." The diverse picture of neuroscience research practice and contemporary work on artificial intelligence presented in this study would not have been possible if so many people had not opened up their labs and taken the time to answer the most dilettante questions.

My most profound intellectual debt, of course, is to the supervisors of my doctoral thesis. Nik Rose and Martina Merz helped me to find my own ways of studying the (neuro)sciences and tirelessly pointed out any weaknesses in my approach. They are both present in this book as advisors and as my science-studies conscience.

Last but not least, I thank my family and friends for their unconditional support over the years. You know who you are.

The research that undergirds this book was supported by Swiss National Science Foundation Grant No. 100016_156786 (Project Title: Machine Love?).

COGNITIVE CODE

A New Cerebral Alley

INFRASTRUCTURAL INTELLIGENCE

Infrastructure is traditionally conceived as substrate – literally structure that organizes something else from below. Recently, however, infrastructure has supposedly become smart, autonomous even, as it increasingly includes technologies fuelled by algorithms that develop on their own in response to ever-growing bodies of data. To be sure, intelligent infrastructure still relies on humans for the tedious informational labour of labelling data. But its inner workings reputedly surpass and even elude us. Whereas information infrastructure is designed to be "ready-at-hand" (Bowker et al. 2010, 99), intelligent infrastructure is destined to get out of hand. The substrate, as it were, develops an intelligent life of its own.

This book traces one of the less visible paths by which infrastructural intelligence is emerging, what I call the *new cerebral alley*. When I conducted ethnographic fieldwork in brain imaging institutes in the United Kingdom and Switzerland from 2009–13, I was not aware that the old link between neuroscience and artificial intelligence (AI) was about to be revived. It seemed to be a thing of the past, most notably an element of mid-century cybernetics, that had been discredited through various "AI winters," which seemed to prove that artificial or machinic intelligence could not exist. However, one of my interviewees, Gareth, the head of a British brain imaging lab, alluded to an analogy between the human brain and planetary-scale information processing infrastructure: "The actual implementation [of cognition] is completely irrelevant. It can be in the Cloud, you know. It doesn't matter." Gareth's lack of concern with the biophysical reality of the brain reflects an attitude that I argue is ingrained in the new epistemologies of brain imaging and cognitive neuroscience. The traditional focus of brain imaging was the structure of the brain – its biological substrate as well as its functional regions – and its goal was to produce an atlas of *the* human brain, an ideal image of the substrate of intelligence. But over the past decade, brain imaging has become increasingly focused on producing data and methods to model brains in silico. And in this process, the biophysical brain recedes into the background of cognitive neuroscience as the algorithmic level of brain function comes to the fore as an infrastructural test bed for intelligence in general.

A CEREBRAL BLIND ALLEY?

I believe we are in an intellectual cul-de-sac, in which we model brains and computers on each other, and so prevent ourselves from having deep insights that would come with new models.

Rodney Brooks, Nature, February 2012

AI researchers should not only immerse themselves in the latest brain research, but also conduct neuroscience experiments to address key questions such as: "How is conceptual knowledge acquired?" Conversely, from a neuroscience perspective, attempting to distil intelligence into an algorithmic construct may prove to be the best path to understanding some of the enduring mysteries of our minds, such as consciousness and dreams.

Demis Hassabis, Nature, February 2012

What I call *infrastructural intelligence* corresponds to views of intelligence in both cognitive neuroscience and related emerging fields such as neuroscience-inspired AI (Hassabis et al. 2017). This is fraught territory. Consider, for instance, a 2012 *Nature* editorial honouring Alan Turing's centennial, which asked experts to comment on the division between neuroscience and artificial intelligence. Rodney Brooks, a former roboticist at the Massachusetts Institute of Technology and the founder of iRobot, criticized the practice of modelling brains and computers on each other, referring to an "intellectual cul-de-sac" (Brooks et al. 2012). His comments were printed under a summary subheading: "Avoid the cerebral blind alley."

Such worry about an apparent cerebral blind alley reflects a long history of debate over the very possibility of machine intelligence. What might variously be called classical, cognitivist, symbolic, or "Good Old Fashioned" (Haugeland 1985, 112) AI, which was dominant from the 1950s to the late 1980s, defined intelligence in terms of rationality, focusing on computing systems designed for high-level abstract reasoning activity, such as playing a game like chess. These disembodied systems were notoriously limited by their lack of real-world applicability. They relied on humans to input and interpret symbolic representations, and they were too brittle to adapt to new problems or circumstances. By the late 1980s, some researchers, Brooks in particular, had instead begun to approach the problems from the perspective of the embodied, environmentally situated sensorimotor skills that subtend the movement and survival of biological systems like insects (Brooks 1989; Brooks and Flynn 1989). For Brooks, the old analogy between brains and computers hides the biological complexity of the brain and mistakenly ignores characteristics of intelligence more basic than abstract reasoning. His new AI focused on sensory-motor response and proprioceptive sense in order to build robots that could process these basic skills into the more complex ability to move around and respond to the environment and others.

Brooks's classic paper "Elephants Don't Play Chess" (1990) communicates loud and clear his view that abstract strategy games have little to teach us about real-world intelligence. An infamous chess match in 1997 underscores the problem. IBM's Deep Blue beat the best human player in the history of chess, Garry Kasparov. Long considered the "*Drosophila* of AI" (e.g., see McCarthy 1990, 227), chess putatively requires an exclusively human combination of skills and capacities, but Deep Blue proved that it was possible to formalize them algorithmically. This outcome, however, failed to mark the beginning of a new age of intelligent machines. The victory instead highlighted AI's narrow focus on its own experimental systems and suggested a loss of sight of the goal of modelling the human brain and intelligence. IBM dismantled Deep Blue's hardware base and discontinued its development. It entered the history of AI as the "playing-chess-against-Garry Kasparov computer" (Ensmenger 2012, 24), a one-hit wonder for a discipline arguably on the skids within a cerebral blind alley. Computer chess became a discipline in its own right, decoupled from investigations of human intelligence.

The sad fate of Deep Blue offered a playful reminder that AI's investments in games should decrease as the real-life application of its techniques increases. As noted by Arthur Samuel (1960), a pioneer in the field of computer gaming, artificial intelligence, and machine learning, "programming computers to play games is but one stage in the development of an understanding of the methods which must be employed for the machine simulation of intellectual behavior" (192).

Now, let me fast-forward to March 2016. At that time, Google DeepMind's family of artificial intelligences had succeeded in matching or surpassing human-level control in many of the infamous Atari 2600 games. But the significance of Google DeepMind had extended beyond competitions between humans and machines in game-playing contexts. Google has implemented deep neural networks in its image search function and in YouTube's user recommendations. They have been used to lower the energy consumption of Google's cloud server farms by 30 per cent, transforming a seemingly passive computing infrastructure into a self-regulating, intelligent infrastructure.

Backed by neuroscience-inspired AI, Google was still more than happy to promote its own game-based confrontation between human and machine. On this occasion, a deep neural net algorithm, AlphaGo, defeated eighteen-time Go world champion Lee Sedol in the Google Deep Mind Challenge. Experts were surprised, if not shocked, by Sedol's loss, despite the fact that AlphaGo had defeated the reigning European Go champion, Fan Hui, in a closed-door match in October 2015. Perhaps the response to Fan Hui's defeat was due to the unofficial nature of the match, or perhaps it was caused by the fact that most European players are considered hopelessly inferior to Korean and Chinese players. Whatever the case, no one, except AlphaGo's developers, expected the algorithm to literally dominate the second-best player in the world.

Part of the difference between the case of Deep Blue and the case of AlphaGo has to do with the game itself. Go occupies special status for programmers and AI researchers at the top of the list of ultimate challenges for AI. Its scope is enormous: Google does not tire of emphasizing that there are more possible positions on a Go board than atoms in the universe, more than the most powerful computers can contemplate. Experts believe it is essentially different from other games mastered by AI. Unlike chess, it cannot be mastered through method alone – but seems to require intuitive sensory skill to play based on how the board looks and feels. Apparently, playing board games and learning in virtual environments are no longer considered parts of a merely preliminary stage in the development of AI.

Of course, not everybody was impressed by AlphaGo's victory. Brooks (2018), for example, drew attention to the old issue of whether machine intelligence is even possible:

> The program had no idea that it was playing a game, that people exist, or that there is two dimensional territory in the real world – it didn't know that a real world exists. So AlphaGo was very different from Lee Sedol who is a living, breathing human who takes care of his existence in the world. I remember seeing someone comment at the time that Lee Sedol was supported by a cup of coffee. And Alpha Go [*sic*] was supported by 200 human engineers. They got it processors in the cloud on which to run, managed software versions, fed AlphaGo the moves (Lee Sedol merely looked at the board with his own two eyes), played AlphaGo's desired moves on the board, rebooted everything when necessary, and generally enabled AlphaGo to play at all. That is not a Super Intelligence, it is a super basket case.

On the one hand, Brooks's characterization of AlphaGo is not inaccurate. Unlike a living, breathing human, AlphaGo is an assemblage of algorithms designed by an army of engineers. It was trained by various Go professionals to recognize and counter human strategies. The algorithm needs human coders and maintainers, as well as the eyes and ears of users who generate data that powers the autonomous evolution of AlphaGo. On the other hand, the characterization misses the planetary-scale orientation of neuroscience-inspired AI. The algorithm integrates the informational labour of an abundant and largely anonymous workforce, but its information processing capacities do not in fact rely on human labour. Rather, this new model of intelligence is *inspired by* experimental neuroscientific models of cognition.

Neuroscience-inspired AI is providing the ground zero for a post-anthropocentric intelligence, one that moves both beyond the goal of modelling the human brain – and the issues of biological complexity involved – and beyond the closely related goal of reproducing human intelligence in machines. Post-anthropocentric intelligence does not have a predetermined substrate, nor does it depend on established boundaries between neuroscience and AI. In an article

entitled "Neuroscience-Inspired Artificial Intelligence" (2017), the
founder of DeepMind, Demis Hassabis, and his co-authors indi-
cate that "biological plausibility is a guide, not a strict requirement"
because what really interests them is "to gain transferable insights
into general mechanisms of brain function, while leaving room to
accommodate the distinctive opportunities and challenges that arise
when building intelligent machines" (245).

NEW METHODS AND EPISTEMOLOGIES

The most significant neuroscientific research technology, functional
magnetic resonance imaging (fMRI), has undergone significant
changes in its roughly thirty-year history. Initially, it aimed to identify and predict structural changes in the brain before they occurred,
and then it was turned into the primary instrument for registering
activations of certain brain structures in response to external cognitive demands. Now, however, it is increasingly used to produce
data that represent complex network interactions – network traffic
– within the brain. The trajectory of the technology of fMRI and the
attendant changes to brain imaging methodology provide an inroad
to the more than subtle shift of focus in cognitive neuroscience that
also undergirds neuroscience-inspired AI.

My fieldwork in brain imaging institutes illuminates conceptual
shifts and methodological twists that, as I show, have paved the way
for a new cerebral alley. The idea of a cerebral blind alley is premised
on the assumption that the substrate of intelligence is the biophysical
brain. This book aims to complicate that assumption by illustrating
how new methods and epistemologies in brain imaging have fuelled
a turn toward what Hassabis and colleagues (2017) call "a systems
neuroscience-level understanding of the brain, namely the algorithms,
architectures, functions and representations it utilizes" (245).

In taking up this shift toward a systems-level understanding of the
brain, this book aims to complexify social scientific and humanistic
treatments of the neurosciences. Scholars of science and technology
studies, the sociology and anthropology of neuroscience, and the
history of science and medicine have long focused on neuroscience's
work on establishing territory for mapping and governing mental
disorder via neurotechnologies. However, as chapter 1 illustrates,
as brain imaging increasingly focuses on the algorithmic level of
brain function, there is an ongoing dematerialization of mental

disorder, intelligence, and cognition. The chapter considers part of this dematerialization through a reconceptualization of the circuit, long figured as the central analytic device of cognitive neuroscience's explanations of the task-specific interactions of functional brain regions. The chapter then looks at the controversy around a 2017 experimental "brain scan" of a legacy MOS 6502 processor – a preliminary climax in the gradual demise of the notion of a hardwired and functionally segregated brain.

The turn toward algorithms and cognitive architectures has been fuelled by and can be observed most significantly in the work of (former) epistemological outsiders – self-identifying "data monkeys," in the words of one of my interviewees, Johan, a postdoctoral researcher employed at a Swiss brain imaging lab – who typically lack a dedicated neuroscience training but command significant expertise in statistics and programming. As discussed in chapter 2, the expertise of methodologists lies not in mapmaking but in the domains of signal processing (i.e., differentiating noise from information) and data analysis (i.e., programming ever more nuanced models of the brain's workings from brain imaging data). Their epistemologies and methodologies, imported from mathematics or engineering disciplines, have gradually changed brain imaging practice and merged into cognitive neuroscience epistemology, resulting in a prevalence of simulation and prediction of cognitive processes over the production of static brain maps.

Chapter 3 focuses on a particularly well-known methodological controversy, the short-lived yet far-reaching voodoo correlations scandal in social neuroscience. This case helps to demonstrate the role of data analysis practices in key epistemological shifts in the social neurosciences. Interesting for this book's purposes is that the main accusations levelled against leading neuroscientists had the potential to cast shadows of doubt over brain imaging methodology in general, but the crisis was quickly contained and resolved. As the chapter discusses, the case of voodoo correlations differs significantly from the current replicability crisis in psychology, which appears to have shaken the discipline to its very core. In the former, the discussion of apparent false positives was bound by the limits of most neuroscientists' knowledge of technical details, and thus the impact of the case on the epistemology of cognitive neuroscience was quite gradual. However, as the chapter argues, the initial appearance of mere corrections to inflated correlations between cognitive stimulus and related brain response now informs the ongoing *infrastructural-ization* of the brain. These seemingly small technical details scaffold the new cerebral alley between cognitive neuroscience and AI.

INFRASTRUCTURES OF THE NEW CEREBRAL ALLEY

As Tung-Hui Hu (2017) has suggested, infrastructure, a seemingly "invisible or cryptic medium" that might be "revealed or decoded," is first and foremost a "speculative medium" (81–2). As information infrastructure, despite its enveloping character, it constantly appears to sink into an invisible background (Bowker et al. 2010) that we feel the urge to visualize and map. Yet the action of obscuring physical structures, what I call *infrastructuralization,* is a voluntary and often purposeful process of rendering invisible the indispensable scaffolds of the technologies that occupy our attention and govern our lives. This is not a matter of conspiracy theorizing; what is at the heart of this process is a reduction of complexity by emphasizing particular aspects and infrastructures at the expense of others.

Smart cities are an excellent example. Billed as habitats of the future and designed to promote novel ways of living in urban environments, they are promoted through images that emphasize the use of natural materials and slick design in the planning of generous parks and waterfront housing. Yet they are engineered as local nodes or interfaces of planetary-scale information and communication infrastructure. At the heart of these infrastructures is the idea of ubiquitous surveillance, just-in-time production and logistics, and infrastructural intelligence – a model of intelligence that reinterprets cognition in terms of distributed information processing. This model aims to approach cognition and intelligence through different sensorial systems and on various scales – the human body and the brain being but an instance of infrastructure.

Infrastructure is not actually "hidden." Nor is the new cerebral alley between neuroscience and AI, as some critics suggest, missing attention to embodied cognition; it simply accentuates other aspects of embodied cognition. An important aspect of information and communication infrastructure is redundancy. The servers that together form the ephemeral Cloud run idle most of the time in the interests of providing constant computing power at all times and in various places at the same time. Virtualization is the process that allows for this infrastructural redundancy to emerge: although we appear to compute on a dedicated server, all we know for sure is that computation occurs. Rather than focusing on the physical infrastructure of computation, we might instead try to understand the processes by which distributed computation is transformed into seemingly homogeneous computing power. This is the focus of computational cognitive neuroscience.

Chapter 4 focuses on a relevant shift of focus within the past decade of cognitive neuroscience research. The 1980s-era discovery of adult neurogenesis and, more generally, brain plasticity suggested that the brain's cognitive infrastructure is to a certain degree redundant. The more recent identification of apparent false negatives in brain imaging data sets demonstrates a brain much more delicately organized than traditional brain maps and atlases would make us believe. By tracing this shift, the chapter unpacks how a loss of confidence in functional segregation and clearly delineated brain regions has contributed to a deterritorialization of networks. It also begins to show how a neuroscientific focus on cerebral geography is being replaced by algorithms and architectures of cognition.

Chapter 5 traces the first evaluation of the European Human Brain Project to further illustrate the rethinking of the biophysical brain. The evaluation was surrounded by controversy over the proposed necessity of detailed simulations of brain structures in silico. According to a loose coalition of neuroscience researchers, this focus on the brain's hardware would be counterproductive to the ostensible goal of achieving an understanding of how the brain processes information. This case helps to show that, although biophysical regions of the brain, such as the amygdala or the posterior-cingulate cortex, do exist, the brain has increasingly come to be described over time as a conglomerate of large-scale networks of neuronal populations that regulate each other. This is a process of infrastructuralization of the brain, one that closely resembles the infrastructuralization of computing in the Cloud.

As chapter 6 discusses, an emerging computational approach to cognitive neuroscience aims to reproduce the time-based interactions of distributed brain structures by means of computational models. This approach follows epistemological guidelines where the work of the "data monkeys" discussed in chapter 2, increasingly come to prominence. Deviating from the path of traditional brain science, cognitive neuroscience is becoming a data science geared toward the simulation of models. The goal is to explain how the brain works,

as well as to devise strategies for correcting maladaptive brain function and for alleviating mental disorder. The attendant, increasing
interest in coding expertise is set to push brain science even further
in this direction by way of certain epistemic tools, such as the programming language Python. In this way, the biophysical workings of
the brain are reconceived by means of techniques and technologies
that provide a platform to think about intelligence independent of a
particular substrate.

POST-ANTHROPOCENTRIC INTELLIGENCE

Cognitive neuroscientists have made first steps in the direction of
developing computational models of cognition and intelligence.
Although I do not claim that cognitive neuroscience is the primary
cause of what I am calling post-anthropocentric intelligence, I do
suggest that the field has played an important role in the infrastructuralization of intelligence and in the development of the kind of
infrastructural intelligence that the stakeholders in cloud infrastructure – whether Microsoft, Facebook, Amazon, or Google – are now
so invested in. Indeed, neuroscience-inspired AI mingles and merges
computational cognitive neuroscience and artificial intelligence
research in pursuit of intelligent infrastructures of the future. The
speculative ambitions of scholars such as Hassabis and colleagues
(2017) involve gaining insights "into some of the deepest and the
most enduring mysteries of the mind, such as the nature of creativity,
dreams, and perhaps one day, even consciousness" (255).

In the speculative spirit of my exploration of the infrastructuralization of intelligence, I have chosen to bookend this Introduction
with a Debriefing rather than attempting to posit conclusions.
Debriefing is a procedure employed in fields such as psychology,
where the participants of a concluded experiment receive a fuller
explanation of the study in which they participated. Debriefings
are necessary since it is rarely possible to disclose the goals of a
study without affecting the behaviour of participants and thus the
results. In many cases, debriefing involves revealing to the research
participant any deception that occurred during the experiment.
However, in rare cases, experiments reveal deceptions on the part
of the experimenters that incite a certain "epistemological dizziness"
in the laboratory (Morawski 2015). For instance, an experiment by
Filevich and colleagues (2013) revealed that in studies of the neural

correlates of feeling free, many tasks administered to participants produced the exact opposite effect: the participants felt constrained when offered a free choice between a number of alternatives. Their study is a reminder that our preconceptions of certain phenomena are often too simplistic, thus blinding us to their polymorphism. In this regard, the Debriefing in this book provides a different reading of the current entanglement between neuroscience and AI – a reading that scrutinizes some of the prevalent preconceptions in the discussion of their relationship.

Although the issues discussed remain attached to the sciences of the brain and psychology, they are inevitably shaped by the technological developments of the times. Neuroscience-inspired AI inveigles us to think of it as an attempt to revive the tired brain-computer metaphor in pursuit of intelligent infrastructure, but its interests suggest a move away from "the brain" and toward the algorithmics of intelligence and cognition. This book is informed by an abiding interest in the phenomenology of mental disorder and involves grappling not only with how the post-anthropocentric intelligence made possible by the processes of infrastructuralization reshapes the way we think about and experience mental disorder but also with how it might provide clues to the impending futures of psychic life.

I

Neuroscience(s)

In a working paper that summarizes results of the Brain, Self and Society project, conducted at the London School of Economics and Political Science, Joelle Abi-Rached (2008) observes that what she calls "the new brain sciences" are to a certain degree disunified. "As in physics, or indeed as in biology in general," she writes, "the brain sciences consists of different cultures and 'sub-cultures' with 'trading zones' between the different platforms." However, "references to 'the brain' – although referring to different conceptual objects – seem to hold together this cosmopolitan background of conceptual, technical, experimental and theoretical varieties" (3–4).

Roughly ten years later, the situation has not changed much. Google's brand of neuroscience-inspired AI, despite representing a new approach to analyzing the brain's workings in silico, prominently integrates older technologies and methodologies such as functional brain imaging with MRI, which still provides the most significant tool for localizing activations of specific brain structures. The central epistemology of brain imaging, however, is about to change within the framework of computational approaches to brain function. Social scientists' concerns about a "neurobiologization" or "cerebralization" of the social subject thus appear increasingly misplaced. At the same time, and counterintuitively, the brain-computer metaphor is losing significance in favour of a novel, engineering-based approach to the workings of the brain, which might revive mid-century understandings of the circuit, which had almost been forgotten.

PSYCHO-BIOLOGY

It would not be false to say that the relationship of the social sciences and the neurosciences has been difficult throughout recent decades. Since the proclamation of the decade of the brain by US president George H.W. Bush in 1990, the neurosciences have attracted a wealth of funding that, according to many observers of neuroscience research, has failed to pay off. In fact, the suspicion that the sciences of the brain could sell old wine in new skins never entirely wore off. At the heart of social scientific critique sits the idea that the sciences of the brain might amount to a (neuro)biologization of qualitative psychological categories and social science theories (Rose 2009). The biggest concern of the critics has accordingly been that the complex figure of the social subject could be reduced to an ensemble of "neurons interconnecting in [a] vast network, discharging in certain patterns modulated by certain chemicals, controlled by thousands of feedback networks" (Gazzaniga 2005, 31).[1]

Said emphasis on the neurobiological correlates of behaviour and the psyche has been criticized by Alain Ehrenberg (2011) as a reductionist naturalism inherent to what he calls the "strong programme," which purportedly "identifies knowledge of the brain, knowledge of self, and knowledge of society" (118). The "affective-cognitive neurosciences," he argues, "have given rise to the idea that social bonds can be explained more effectively in terms of neurological foundations than from a sociological perspective" (119). This anxiety about the looming emergence of a brain-based social subject has been fuelled by the institutionalization of "social neuroscience," which interweaves distinct theories and methodologies in pursuit of identifying the cerebral correlates of social cognition and explaining the development of the contemporary human brain as a result of social interaction.[2]

Latching onto evolutionary arguments (e.g., see Emery 2012), social neuroscientists typically conceive sociality as an anthropological constant and an individual capacity, suggesting, as historians of psychology Nikolas Rose and Joelle Abi-Rached (2013) observe, "that the brain has evolved to favour a certain type of sociality manifested in all the interactions between persons and groups that come naturally to humans in our social lives" (163). "Normal" brain development and healthy brain function are accordingly deemed characteristic for the social individual to the same degree as a "healthy" social upbringing supposedly results in a smoothly functioning brain. From this perspective, dysfunctional social behaviour can be considered a sign of major psychiatric syndromes, and the process of socialization is spelled out as becoming oneself in terms of the brain (Ehrenberg 2009).

According to science historian Fernando Vidal, the contemporary pertinence of everything "neuro" is in fact part of a much longer history of scientific explorations of the subject and extends John Locke's substance-indifferent individualism into contemporary fields of neurobiological reasoning (Vidal 2009b; see also Ortega and Vidal 2007, 2011; Vidal 2011). Vidal (2009a) argues that beginning with the science of phrenology, "brainhood" reconstituted the anthropological figure of modernity based on theories of the Viennese physician Franz Joseph Gall, who hypothesized that the brain must be the organ of the mind and that the faculties of the mind ought to have a distinct location within the brain. Brain imaging, oftentimes regarded as phrenology's contemporary complement or a "neo-phrenology," has now extended the reach of the argument by making the social subject readily available through noninvasive investigations of brain function by means of brain imaging technologies (Hagner and Borck 2001).[3]

The suspicion that these images might exercise a "seductive allure" on their beholders (Weisberg et al. 2008) cannot be confirmed.[4] Instead, the team of psychologist Deena Weisberg has found that neuroscientific explanations possess a certain "reductive allure" that appears to be strongest for the pairing of brain science and psychological theories (Hopkins, Weisberg, and Taylor 2016; Weisberg, Taylor, and Hopkins 2015). Whereas volunteers who participated in the study judged sociological theories largely independent of additional neuroscientific information, they found psychological theories to be more credible when neuroscientific addendums were provided. The fact that this effect remained stable even when the neuroscience explanation was entirely nonsensical suggests that psychological theories have lost ground to and are in danger of being undermined by the "other" science of the mind.[5] Neuroscience appears to have emerged as a "psychology plus," or the realization of psychology's potential by other means, as in substituting abstract categories with "a vocabulary that originates from neurobiological reality rather than human invention" (Alcalá-López et al. 2017, 2227).

An important aspect of the criticism leveraged against the brain sciences has accordingly been the notion that their expertise in measuring physiological processes and analyzing biological substrates is typically outweighed by impoverished understandings of the psyche and sociality. Historian Nancy D. Campbell (2010), for instance, contends that social factors are allowed into neuroscience only to the extent that they remain "reductive and abstract, rather than concrete and substantive" (91), and anthropologist Emily Martin (2010) suggests that "we are seeing the effects of a form of [neuroscientific] reduction that is likely to impoverish the richness of human social life" (369).

The methodology of brain mapping, for instance, has been criticized for supporting a "neo-phrenology" (Uttal 2001; for a history of the term in science, see Cornel 2017) that has revived nineteenth-century efforts to locate the "organs" of the mind within the brain. Contemporary brain mapping is based on identifying brain regions and their cognitive functions; in experiments, these regions are triggered with certain cognitive stimuli or mental tasks. This stimulus-response approach has been criticized for latching onto a specific experimental paradigm rooted in mid-twentieth-century psychologies. The social neurosciences in particular have inherited a methodological preference for experiments conducted with volunteers in the controlled environment of the laboratory, where standardized experimental tasks substitute for the "real" social situation (Easton and Emery 2012). With their bodies suspended in the scanner and deprived of the presence of other humans, volunteers encounter sociality primarily through "sociality-inducing" stimuli: images, motion pictures, or sterile interactions with researchers.

In his ethnographic study of brain imaging in Britain, Simon Cohn (2008) observes a continuity of basic epistemological assumptions based on localization and materialism, where what social neuroscience "describes as 'social' becomes merely an extension of the same restricted notions of human nature that informed earlier periods of behaviourism" (101). Volunteers or data bodies are configured in the scanner room in order to observe how the biological substrate reacts to "social situations," which are again standardized, "instant" versions of what we encounter in daily life designed in pursuit of comparability between individual experiments.[6] The experimental lineage of brain imaging, so the argument goes, qualifies the sciences of the social brain as a social psychology that stands on (neuro) biological feet (see also Matsusall 2012; Matusall, Kaufmann, and Christen 2011).[7]

"Simply put, then, what is placed under the scanner is not the psyche as such but, rather, psychological theory itself," Jan De Vos argues in his recent book *The Metamorphoses of the Brain: Neurologisation and Its Discontents* (2016), cautioning against the "tautological risk" that "psychology is supposed to underpin neurological research while the latter is more and more evoked as the final proof of the scientific validity of the psychological theories themselves" (34f).[8] His statement is reminiscent of the many other critical accounts of neuroscientific practice to emerge throughout the past decade that have maintained, for instance, that "the rhetoric within neuroscience and biological psychiatry ... reflects the conviction that future advancements in neuroscience will ensure the displacement of several psychiatric practices – including psychodynamic, social and cultural psychiatry – by biological approaches" (Choudhury, Nagel, and Slaby 2009, 71).

Whether tautological risk or entire displacement, the relationship between the purportedly "qualitative" social science of the mind and the "quantitative" biological sciences of the brain has certainly been highly complicated and has (deservedly) received a lot of attention. Depending on where you stand, it seems fair to argue either that the neurosciences have come to undermine psychological theories or that the neurosciences have come to the rescue by augmenting psychological theories with "hard" scientific facts in the realm of both analysis and therapy (Beaulieu 2003). Supported by neuroscientific empiricisms, psychological categories remain valid or might even gain significance; the qualitative approach to the human mind characteristic of psychology, however, appears to be thwarted by the neurobiological register.

CIRCUITS

In contrast to the majority of neuroscience-critical scholars, sociologist Des Fitzgerald and geographer Felicity Callard (2015) vouch for an experimental entanglement of the social and neurosciences, maintaining throughout a number of collaboratively authored texts that there is much more to experimental near or quasi realities in cognitive neuroscience than a mere "cerebralization" or "(neuro)biologization" of psychology's subject. Since functional brain imaging, for instance, etches together "local politics, de-oxygenated blood, sick bodies, nuclear physics, and the clinical gaze" (11), the cognitive neurosciences' preferred empiricism confronts psychology and the social sciences with an opportunity to revaluate their own experimental categories and assumptions – not in disavowal of neuroscientific knowledge as fabricated but through a well-due appreciation of the technicalities that also undergird "qualitative," psychological, and social science understandings of the social.

Nikolas Rose (2013, 2019), as much as Fitzgerald and Callard, emphasizes the neuroscientific project's potential to link biology and psychology in ever new ways and thus to correct a blind spot of twentieth-century social science, namely the various ways that what we conceive as social and cultural complexities have been technologically mediated.[9]

Fitzgerald's book *Tracing Autism: Uncertainty, Ambiguity and the Affective Labor of Neuroscience* (2017) provides an example of such a productive iteration of qualitative categories, where the technicality of neuroscience accounts of sociality and the psyche is not conceived as necessarily reductionist. The sociologist observes "that the specific, dry, and technical issues about the objective make-up of autism that skate endlessly across the top of these accounts are *not* simply a way to avoid talking about love; they are there, in fact, precisely to explain it" (141, original emphasis). It is the final sentence of a section titled "Objects of Love" (134), where Fitzgerald grapples with the significance of what he initially conceived as unbearably technical understandings of autism but eventually came to acknowledge as the source of an affective relationship between autism neuroscientists and children diagnosed with autism spectrum disorder (ASD).

Fitzgerald (2017) observes that the notion of circuits in the brain provides the means to analyze and communicate a form of "difference" that is extraordinarily hard to grasp, particularly since autistic children tend to elude unwanted social interaction. Qualitative accounts of ASD often emphasize the symptomatic withdrawal from social life instead of investigating ASD as yet another form of social behaviour that we simply have trouble understanding.

However, the "dry, and technical issues" that derive from representing difference based on a malfunction of circuits in the brain allow for an affective relationship to emerge; deficient social behaviour turns into otherness that needs to be engaged and at the same time gives reason to reconceive understandings of social behaviour put forward by the social sciences. What if the affective labour of autism neuroscientists were indeed impossible without these unbearably technical accounts of ASD? What if thinking about ASD as an alternative configuration of circuits in the brain could contribute to embracing the condition instead of categorizing it as a disorder or disease?[10]

By embracing the concept of the circuit, Fitzgerald (2017) reminds his readers of the fact that circuits have always played an important part in the epistemologies of cognitive neuroscience. Out of an invisible background, they have governed how neuroscientists imagine their research object, experimentally put it to the test, and transform signals from the scanner into data of brain activation. As abstract models of network interactions between groups of neurons, circuits tend to float in and out of neuroscientists' attention, always part of the experimental constellation yet rarely investigated themselves.

Joe Dumit (2016) observes that circuits have come to be inextricably entwined with the epistemology of cognitive neuroscience through a diagrammatic practice that is rooted in mid-century efforts to model cognition in between the brain and the computer. Ever since, flowcharts have acted as hypotheses of brain imaging studies, mediating between anatomical maps and functional circuits. "In other words, every single box in the scribbled flowcharts was meant to correspond to a particular area of the brain, but also to represent the execution of a particular 'mental step'" (221). Circuits in the brain would hence represent material instances of the (deservedly) much-criticized brain-computer metaphor – in contrast to the intentions of the originators of flowcharts. The drawing practices of Hermann Heine Goldstine and John von Neumann (1947) and Warren McCulloch (1945) (see also Pitts and McCulloch 1988), for instance, had been part of an experimental epistemology devised to keep track of the incredible liveliness of mechanisms instead of fantasizing hardwired cognition into existence.

Dumit's observations show that circuits must not necessarily be conceived as hardwired; rather, they acquired said quality within the field of cognitive neuroscience by way of a certain brain imaging methodology, which is currently "under review." In a recent experiment, two researchers working at the crossroads of data science and neuroscience tested said assumption and applied commonly accepted neuroscience methodology to an accurate simulator of the MOS 6502 microchip (see figure 1.1), which had once powered the

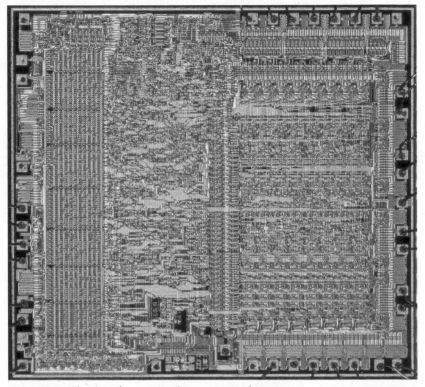

Figure 1.1 The die of an original MOS 6502 chip.

first Apple computer as well as the Commodore 64 and Atari VCS
(Jonas and Kording 2017).

Running three legacy video games as example behaviours of
the circuit, Eric Jonas and Konrad Kording (2017) analyzed all
hardwired connections, the effects on the chip's behaviour when
individual transistors were destroyed, tuning curves, the joint statistics across transistors, local activities, the estimated connections,
and whole "brain" recordings based on standard techniques that are
popular in the field of neuroscience. Despite having experimental
omniscience – due to the fact that we know exactly how a processor
works – and a set of data that would in the neuroscience world be
worth millions of dollars, Jonas and Kording did not even come
close to explaining how computation arose from the workings of the
hardware (see figure 1.2). Can a neuroscientist understand a microprocessor? The answer is apparently no.

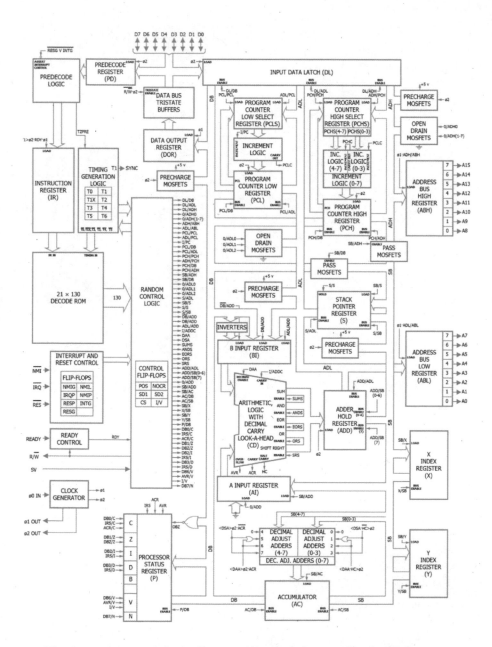

Figure 1.2 MOS versus brain diagram. Above is a diagram that details the hierarchical organization of the MOS 6502 chip. For each functional module, it is known how the output depends on the input. Opposite is a diagram adapted from the classic Felleman and van Essen diagram of the primate visual system.

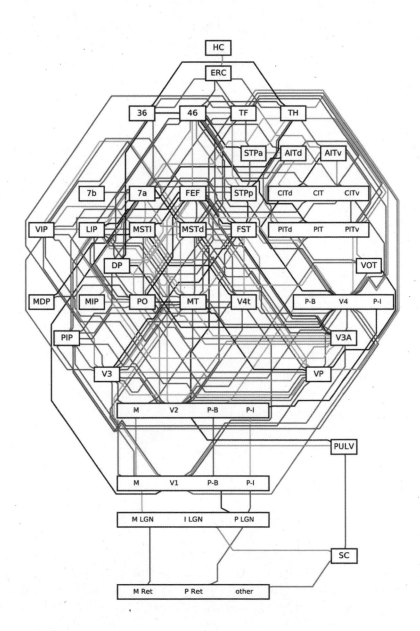

The functional areas depicted in this diagram are of uncertain nature; according to Jonas and Kording (2017), "there is extensive debate about the ideal way of dividing the brain into functional areas. Moreover, we currently have little of an understanding how each area's outputs depend on its inputs" (fig. 13).

Jonas had come up with the idea for the experiment after stumbling upon "microchip archaeologists" – the Visual 6502 project[11] – who had painstakingly reconstructed the classic MOS 6502 chip by employing techniques that neuroscientists routinely make use of to understand the brain: imaging the circuitry within, labelling different regions, identifying connections, and building an exact software simulator of the chip. "It shocked me that the exact same techniques were being used by these retro-computing enthusiasts," Jonas told Ed Yong (2016), a staff writer for *The Atlantic*. "It made me think that the analogy is incredibly strong."

What struck me is that many commenters, including Yong, interpreted this analogy as one between the chip and the brain, despite the fact that Jonas placed an emphasis on the very techniques by which "microchip archaeologists" and neuroscientists study the substance of cognition and intelligence either in the brain or in silico. The chip, or the circuit, hence figures as a speculative medium by which brain mapping in cognitive neuroscience is reconceived as a form of archaeology, if you will, in that it targets identifying and excavating the brain's hardwired infrastructure. However, the analogy drawn by the experiment is much more nuanced and rather critical of the brain-computer metaphor. At the heart of the experiment sits a methodological question: if we could map every functional unit and connection in the brain and track their activity, would we have the tools to make sense of what we discovered? Or, put differently, can the brain be conceived as an assemblage of hardwired circuits?

Testing neuroscientific techniques on a physical circuit was thus turned into an epistemological critique of the analogy between brains and hardwired circuits. I read Jonas and Kording's (2017) paper not as geared to reconceiving the brain and the computer but as an attempt to rethink computation and cognition by turning the wide gap between the brain and the Cloud into a liminal space of curiosity-driven exploration. Indeed, a loose coalition of neuroscientists praised the efforts of Jonas and Kording as a wake-up call for cognitive neuroscience, among them Steve Fleming (2016) of University College London's Wellcome Centre for Human Neuroimaging, who aptly summarized an emerging sentiment within the field of the brain sciences within a short post on his blog *The Elusive Self*. If a better understanding of computation cannot be derived from a data set of hardware activity valued at millions of dollars that in the real neuroscience world would be generated with cutting-edge neuroscience methodology in a controlled environment, why do we expect that a comprehensive understanding of human consciousness will emerge from generating even more data about the connections between populations of neurons within the brain?

Fleming (2016) maintains that "we might be able to understand a particular computation or psychological function to a sufficient degree to effectively 'fix' it when it goes wrong without understanding its implementation (equivalent to debugging the instructions that run Donkey Kong, without knowledge of the underlying microprocessor)." What sounds like a modest proposal has a radical edge to it: can the neurosciences abstain from producing ever more data about probably existing circuits in the brain and instead delve deeper into the full complexity of existing data sets? If, as Dumit (2016) argues, brain imaging demonstrated the existence of hardwired circuits in the brain because data was analyzed by means of "algorithms with built-in assumptions that circuits were present" (221), what would it mean for cognitive neuroscience and thus for our understanding of human cognition if novel methods and algorithms were to cast doubt on the existence of hardwired circuits in the brain?

Fleming (2016) describes Jonas and Kording's paper as a reality test for cognitive neuroscience that is based on brain imaging, but it does, of course, also promote a specific approach to neuroscience that is directly related to an emerging, neuroscience-inspired AI. In fact, Kording is a frequent collaborator of Google DeepMind researchers and has recently vouched for integrating the cognitive neurosciences' focus on the detailed implementation of computation with a perspective from machine learning, which would eschew precisely designed codes, dynamics, or circuits in favour of the gradual optimization of initially very simple architectures (Marblestone, Wayne, and Kording 2016). Within such approaches, the circuit as a model of the brain's functional organization is not entirely dismissed; rather, it has been updated to fit the mould of neuroscience-inspired AI.

One of my interviewees, Nik, then the head of physics in a British brain imaging lab, explained what an attention to cognitive algorithms means for the brain-computer metaphor and the neuroscientific project in general: "If you have an algorithm in a computer, this is a very abstract description of what the computer executing an algorithm is doing, but if you have the algorithm, you know exactly what the computer is doing, and that's how we control computers

– even though you have no idea how this works in detail with the transistors and capacitors at the level of electronics, right?"

Instead of mapping the biophysical components of the brain, a computational approach to cognitive neuroscience involves focusing on the rule-based processes that govern information processing in the brain. In pursuit of neuroscience-inspired AI, Google DeepMind has been very outspoken about the fact that the design of its algorithms was inspired by its CEO's research on episodic memory and imagining the future. Demis Hassabis has identified episodic memory, or the capacity to travel back in time and remember subjective experiences in vivid detail, as a key mechanism in the simulation of future events and thus in the development of solutions to problems that might not yet have occurred (e.g., see Hassabis et al. 2007; Hassabis et al. 2014).[12] Episodic memory is a cornerstone of neuroscience-inspired AI, but it is just one aspect of a more general interest in various forms of memory, which may render learning more efficient, sustainable, and ideally, lossless.

In this regard, the sciences of the brain are to provide a rich source of inspiration for novel cognitive architectures that are plastic enough to be continuously moulded and simultaneously too robust to bend under conditions of information overload. The neurosciences are called upon to generate a "ground truth" for algorithmic developments: a level of cognitive capacities that algorithmic systems should at least be able to reproduce, if not surpass. At the same time, these systems are hypotheses of how cognition might be implemented in the brain. And "[i]f a known algorithm is subsequently found to be implemented in the brain, then that is strong support for its plausibility as an integral component of an overall general intelligence system" (Hassabis et al. 2017, 245).

In other words, the brain is understood as a sort of analogue, experimental system that artificial intelligence should not hesitate to exploit. In this context, the neurosciences fulfil a precisely defined function, and Google DeepMind's cast of neuroscience-savvy AI researchers do not shy away from providing recommendations for what the brain sciences should prioritize in the near future.

Hassabis and colleagues (2017) show interest in a level of neuroscience research where the brain is figured as providing no more than an "infrastructure" for abstract concepts or systems such as working memory, which characteristically hover in between computing and psychology. This level may be described, as Google DeepMind researchers tend to do, as a systems level of neuroscientific inquiry; it might, however, just as well be figured as computational (cognitive) or theoretical neuroscience. In any case, the neuroscience that undergirds the novel cerebral alley seeks "to better understand the brain's workings at the algorithmic level – the representations and processes that the brain uses to portray the world around us" (Hassabis, in Brooks et al. 2012, 463).

The neuroscience of neuroscience-inspired AI is remarkably different from the brain sciences that have been investigated by scholars of the humanities and social sciences. Most significantly, the algorithmic level of neuroscience is indifferent to the exact implementation of the underlying processes. In other words, the "precise mechanisms by which this is physically realized in a biological substrate are less relevant here" (Hassabis et al. 2017, 245). With the fading relevance of biology comes a reorientation toward engineering perspectives and the informational labour of methodologists, who are tasked with gradually merging the concept of hardwired circuits typical of twentieth-century cerebral geography into models of self-actualizing algorithms that resemble those of early cyberneticists.

What follows is an investigation into the lives and practices of the stakeholders of statistics and computational methods in brain imaging. These self-identifying "data monkeys" (Johan) are typically not conceived as the originators of cognitive neuroscience epistemologies; instead, they embrace the role of outsiders, who another of my interviewees, Rita, a senior researcher in a British brain imaging lab, said are "kind of largely outside the domain of understanding what the brain is doing." As the following chapters suggest, however, these humble gestures do not always paint an accurate picture of the distribution of epistemic authority throughout the field. "Data monkeys" command skills that are extremely valuable to "real neuroscientists" and have ample opportunity to merge their own goals and interests into the traditional epistemologies of the brain sciences and thus (re) imagine in practice the discipline's central research object. To put it in the words of Hassabis and colleagues (2017), from "an engineering perspective, what works is ultimately all that matters" (245).

Informational Labour

Rita told me that "the lab benches are essentially the scanners down-stairs, but we only go there if we *really* need to scan someone." She continued by explaining how methodologists stand out from the crowd of "real neuroscientists." When I asked her whether there was a gap between experimental work and methods work, she replied,

> Yeah, there's a gap on a number of levels. The people that work in these fields are different in neuroscience. So in this building, there'll be computer scientists and mathematicians and engineers, so they tend to work on developing methodologies and statistical techniques. You obviously need to know a degree of neurophys-iology and biochemistry, trying to understand actually what you are looking at, but a lot of it is model development, statistical inference ... But these are kind of areas of mathematics ... So you're kind of largely outside the domain of understanding what the brain is doing. You're more in the domain of understanding what the signal tells you as a methodologist. And so we tend to be more computational people. And then ... the people that come through the domain of experimental work are particularly interested in the biology, so clinicians, psychiatrists, neurologists, psychologists, and again neuroscientists.

This distinction between experimenters and methodologists is highly prevalent throughout the field of cognitive neuroscience. On one side are "real neuroscientists" focused on scanning volunteers' brains to identify the neural correlates of the cognitive functions or pathological malfunctions of said brain structures; on the other

side are "computational people" working with statistical or computational methods toward the improvement of signals received
from fMRI machines and the refinement of data models that relate
experimental tasks to measured brain activity. Methodologists work
not in facilities that resemble the classic wet labs of biology but in
office-style buildings that differ from those of "insurance companies
merely in that the computers are a bit more fancy" (Nik).

Anthropologist-turned-neuroscientist Andreas Roepstorff (2002),
argues that the architecture of brain imaging institutes is indeed
instructive as regards the distribution of specific tasks throughout the
process of transforming brains into data. During an ethnography of a
British brain imaging institute, Roepstorff observes that the organization of St John's House in London follows the top-down hierarchical
model of a typical Georgian house, where goods came in on the ground
floor, were brought down to the kitchen in order to be processed, and
eventually made their way up to be consumed in the chambers of the
residents. In the case of the brain imaging institute, brains enter the
building on the ground floor to be guided down to the scanner room,
where they take part in experiments and provide signals that the scanner registers. Brains, as brain imaging data, then make their way up
to be processed and turned into information, before reappearing as
results to be discussed with directors and principal investigators in
institutional meetings on the top floor.

The way Roepstorff describes the movement of data through the
house emphasizes the quasi-magical transformation of brains into
"immutable mobiles," as Bruno Latour (2012, 237) would have
it. Although he mentions the predetermined oscillations between
steps in the analysis on the intermediate floors, the journey of data
from the basement to the top and out into the world through the
chimney appears surprisingly streamlined and obscures the variety
of processes that data undergo, or the informational labour that
methodologists perform. Brain imaging in cognitive neuroscience
involves many "cooks," and those who put the finishing touch on the
plates are often *not* the chefs. Although they are employed to handle
and manage the intricacies of "raw data," methodologists are often
considered mere *coup de main*; in this chapter, however, I show that
they have ample opportunity to rain on the experimenters' parade.[1]

What follows is an investigation into the lives and practices of the
stakeholders of statistics and computational methods in brain imaging. These self-identifying "data monkeys" (Johan) are typically not

conceived as the originators of methodological thought; instead, as I show in the following sections, they typically embrace the role of outsiders who are "kind of largely outside the domain of understanding what the brain is doing" (Rita) and whose work, as explained by one of my interviewees, Stan, the head of analysis in a British brain imaging lab, is "getting sucked into ... neuroscience" rather than being designed for investigating the secrets of the brain. However, as this chapter suggests, said humble gestures do not always do justice to the actual distributions of epistemic authority within the field. Being assigned to the esoteric circles of cognitive neuroscience or adopting the position of an outsider within those circles implies the command of skills that, being extremely valuable to the field, guarantee a certain influence on the (re)imagination of a discipline's research objects. As sociologist Mikaela Sundberg (2006) states, "the significance of boundary work lies not only in the position of a putative boundary but also in the kind of order it implies" (59).

The next section introduces methodologists' take on the brain, which is typically mediated by statistical models and an abundance of data. In the middle section, I proceed with a more detailed account of their methodological work, which transcends the boundaries of disciplines and research fields. I show how software suites such as MATLAB provide the means, for instance, to turn the results of experiments in the scanner into the raw material of experiments with statistics and to bridge the epistemological gap between populations of neurons and celestial bodies. As discussed in the final section, "coding" thus turns into an epistemic practice that reimagines the brain in the context of data science and gradually disentangles it from its biophysical reality.

OUTSIDERS

Stan said that "brain imaging is a team sport. You are collaborating with medical doctors, physicists, biologists, engineers, computer scientists, and really with psychologists or neuropsychologists and neurologists."[2] Trained as a physicist and engineer who had once specialized in computer vision, yet well established within the field through related work, such as his contributions as an editor to major journals, he addressed the position of a data analyst who ended up "getting sucked into doing bits of neuroscience." He explained that collaborative interactions with neuroscientists tend to be scattered and not necessarily of high durability. "I could be involved in a project from the beginning in terms of experimental design through data analysis and writing the paper," he said, but also constituting his daily reality were "lots of very small interactions to do with questions of statistics." As I was told by another interviewee, Dejan, a methodologist and the head of a Swiss medical image processing lab, "there are so many aspects of statistics and tuning ... that we do a little bit of consulting also." He laughed at the thought but was clearly not joking.

The business of "consulting," however, remains a tricky one: when experimentalists and methodologists speak about the brain, they often approach "the object" from two opposite ends of a spectrum that defines how different the brain can look in the eyes of "cognitive neuroscientists." Whereas experimentalists are occupied with producing significant activations within the brain, methodologists focus on (modelling) the clouds of "raw data" that experimentalists generate. In order for data to be raw, however, if first needs to be processed by an array of statistical devices that methodologists maintain but rarely make use of themselves.[3] "Most of the people in the methods group [...] never collect their own data, so they never have reason to go through the whole process of collecting data and then using the software to analyze it," Stan told me. "So that's a bit odd in a way because we're making that software but we never actually use it. And that's a slight caricature, but the people that use the software most are the empirical neuroscientists who collect the data and want to find a particular thing."

According to Stan, interaction between "real neuroscientists" and methodologists is often mediated by the software that automatically eliminates the noise introduced by experimenters and volunteers. Indeed, many self-identifying methodologists frame their relationship to the brain sciences as one characterized by exteriority and distance from a core of "real neuroscientists," who are conceived to be in charge of defining the general lines of inquiry for the field (see also Roepstorff 2001). Due to an apparently uneven distribution of expertise throughout most institutes, a certain collaborative promiscuity (Merz 2006, 102, 105) and five-minute interactions in the corridors represent the norm rather than an exception for many methodologists, who often merely approach the brain through "the signal" (Rita) or work on its representations in their capacity as "data monkeys." In their own interpretation of interdisciplinary collaboration, methodologists typically figure as computational clerks of sorts, being unlikely pieces of information infrastructure who are replaced by software tools in daily business and intervene primarily to maintain statistical devices or to shock principal investigators with the odd idea about how to visualize the workings of the brain in an entirely novel way.[4]

This is to say that the opposition between methodologists and "real neuroscientists" is not necessarily as tangible as it might appear to be in said accounts of interdisciplinary cooperation; some self-identifying methodologists do have important roles within their institutes, including as directors of powerful methods groups or as "converts" who can no longer be distinguished from "real neuroscientists." The distinction is primarily upheld to make claims in the fight for epistemic authority and to draw lines in the sand that provide a common ground for physicists, psychologists, biologists, engineers, and computer scientists. As the following example shows, the most significant boundary in this respect appears to be defined in practice or, put differently, by the sort of experiments that the surprisingly distinct and self-contained groups of researchers tend to prefer.

"I quite like my computer and programming and running different kinds of analysis. So it suits me to only see people from the general public every twelve months, like that's enough," Rita told me when I asked her about her take on brain imaging experiments. This is to say that she was not particularly fond of scanning in general and even less so if she had to put "people from the general public" into the scanner. Johan confirmed that methodologists tend to avoid recruiting subjects from outside of the university and added that doctoral students, with their "university brains," were by far be the best subjects to use when doing work on methods: "They know how to behave and produce less artefacts in the data."

Instead of having to trick "people from the general public" into certain behaviours, experimenters can expect that education and familiarity with the lab setup will result in resilient correlations between the volunteer and a targeted brain response. "Because if people are not aware that this whole field of science exists, you know, [and] you bring them downstairs and you're scanning them, there's connotations of hospitals and things wrong with you, you know? These are some interpretations that people may be bringing into a task and you don't want that." Whereas experimental subjects are carefully selected in a clinical setting to provide a sort of model organism, "data bodies" recruited by methodologists might be picked in anticipation of a good performance and due to their availability "in the middle of the day" (Johan).

As the term "data bodies" suggests, methodologists are ultimately not interested in volunteers who, for instance, exhibit specific pathologies but are interested in data sets that derive from an undisturbed scan and a preferably "clean" signal, which may well be noisy if said noise originates in the brain activity of the volunteer. In fact, a noisier signal might provide clues about how the signal could ultimately be improved. Another of my interviewees, Yann, a professor at a university in southwestern Switzerland who was trained as a physicist and specialized in methodologies for signal processing, had worked on the improvement of data acquisition methodology in astrophysics before lending his expertise to biomedical imaging. He explained the difficulties that interdisciplinary work entails, providing insights into an ongoing process of methodological reconfiguration:

> You need to perform quite a lot of work to get the experts in that community to understand what you do, to make them understand the concepts, so that they understand it can be interesting for them because they've been working in that field more than fifty years. Of course, they are experts in what they do and they do very well what they do, so they may be resistant to just ideas flying around if there's no proof that at the end of the day it can really be applied to their specific problem with their specific constraints and that it can solve the challenges which are their challenges. So this idea of bridging the gap between the communities, that is what is promoted everywhere, right? Interdisciplinarity. But when you are really in the game, you understand that it's not something that is easy. People are not always directly open to these ideas – to newcomers and new ideas which can change your perspective and things and which make you lose a bit your stability – because you know how to analyze your images today [and] you know how they have been acquired, and radiologists know how to study their images and they know which kind of artefacts they can have from the imaging, and they don't care about these. And if we give them new images, then other artefacts could be there due to the processing, so they don't like them.

A sense of despair filled the air when Yann lamented the strength and stability of "esoteric circles" (Fleck 1979, 105) within brain imaging. He never mentioned the "cross-cultural exploration[s]," "floundering expeditions," and "excursions to the 'frontiers' of knowledge" that Julie Klein (1994, 77) found characteristic of collaborations between researchers with different backgrounds. Instead, Yann clearly took the perspective of a stranger to the field, determined to find his place among the "natives." However, he did not seem to be too concerned about his foreignness and proved resistant to the traditional epistemologies of brain imaging; although a physicist, he was interested in the mathematical commonalities between brains and celestial bodies in outer space. Interdisciplinary research was his mantra, and he had arranged his entire scientific life to facilitate such research by establishing a working group that typically oscillated between biomedical imaging and astrophysics.

In Yann's practice, images figured primarily as models to "render" the world of the brain "visible, tangible, and workable" (Myers 2015, x).[5] In general, methodologists employ specific data visualizations to analyze brain function and experiment with models simultaneously, and they make use of classic brain images mainly "because you need to communicate to the neuroscientific community, to the medical doctors and these kind of people, who might not understand or care about the value of more data-oriented visualizations" (Johan). If such visualizations appear discontinuous to, and thus threaten to break, paradigmatic empiricisms, they risk being rejected and tellingly relegated to the status of (data) visualizations by the esoteric circles of the field. However faithful they may be to that which is regarded as the natural appearance of the "thing," visualizations tend to fulfil their purpose primarily if they latch onto an established empiricism and thereby provide the phenomenon with an "image." This is to say that have we reached a point where the exact "purpose" of imaging or visualizing must be more clearly defined.

In the case above, Yann strategically disidentified with a tradition that defines imaging as an extension of radiological empiricisms. By positioning himself in opposition to a traditional brain imaging culture – "*their* problems, *their* challenges, *their* images" – Yann addressed MRI as part of a longstanding medical imaging culture that dates back to the early 1980s. MRI scanners had not always been classified as radiological "imaging" technologies. In a book on technological change in medicine, Stuart Blume (1992) cites a 1983 report by the American Hospital Association that proves the epistemologi-cal identity of MRI scanners was still entirely undecided in the early 1980s: "Some believe that because NMR [nuclear magnetic resonance] is more than just an imaging device, it requires closer cooperation between spectroscopists, pathologists, biochemists, and other physi-cians outside the radiological field ... Some believe nuclear medicine physicians and internists may be best equipped to understand NMR because of their experience in metabolic physiology" (219).

Especially in North America, however, where the radiological societies were exceptionally powerful at that time, radiologists successfully demanded that MRI scanners be placed in hospitals' radiology departments, and they managed to significantly influ-ence the design and hence also the output format of scanners (Joyce 2008).[6] The technology formerly known as "nuclear magnetic res-onance," which displayed its measurements as arrays of numbers, turned into MRI and was configured to support the transformation of numbers into visual information, which did above all aesthetically reference the sort of images that radiologists had grown used to: grey-scale, X-ray-like representations of the body's "interior."

Yann, by contrast, handled the brain, or rather brain activity, in the form of a noisy signal or data and reverted to visualizations primar-ily to inspect the effects of changes that he made at the level of code. Having observed that "radio astronomy and MR imaging actually share very strong similarities in terms of the mathematics of the acqui-sition of the signal, even though the physics is very different," Yann then provided clues about how the relationship between brains and outer space should be conceived. Here, outer space is not regarded as a physical reality – as an "outer place" (Messeri 2016, 23) – that is full of smaller, comparable worlds of which the brains biophysi-cal reality is one. Instead, it is the signal provided by space telescopes and fMRI scanners that ultimately exhibits the celestial bodies and neuronal populations that tend to hide within a dark and noisy back-ground. Here, of interest is a similarity between cognitive neuroscience and astrophysics that is based on recovering "significant" events from noisy data by means of statistical devices. Here, the analogy is not one of physical nature but derives from the ecosystem of data science that mathematical models and computational platforms provide. They stand for an outsider perspective on the workings of the brain, as it were, that seemingly deviates from the focus of cognitive neuroscience yet significantly influences the appearance of its research object.

MATHS

In *"Raw Data" Is an Oxymoron*, Lisa Gitelman and Virginia Jackson (2013) observe that "every discipline or disciplinary institution has its own norms and standards for the imagination of data, just as every field has its accepted methodologies and its evolved structures of practice" (3). The latter might, however, not always conform with the former. Methodological thought aims to ensure that techniques and technologies employed in experiments do not interfere and cloud the judgment of the experimenter. In an intriguing attempt at a sociology of the neuroscientific epistemology of love, Gabriel Abend (2018) identifies and dissects choices and assumptions that neuroscientists must make to elicit love in the scanner room and hence provides a novel, "messy" account of what love is experimentally. Typically elicited through images of loved ones, love is defined as loving or being in love with someone. At the same time, it is conceived as possibly occurring in the absence of loved ones, being a capacity of the individual that can be triggered, measured via the changing magnetic properties of cerebral blood flow, modelled as a temporally discrete event, and represented as temporal and spatial patterns of activity in and between distinct neuronal populations.

To capture love as an activation of the circuits of love in the brain, it must be isolated and disentangled from all things purely experimental: the choices and assumptions made by neuroscientists but also the ambiguities and uncertainties that characterize the process of defining love for experimental purposes. To capture "raw love," as it were, experimental data must paradoxically be cooked. This time, however, the determinants of cooking derive from a culture that surrounds the production of data: to eliminate inexplicable variance and to ensure comparability, signals have to be cleaned from the traces of the experiment, liberated from the "noise" generated by the scanner, and "normalized" by means of comparable sets of data. Data is cooked to manage the potentially poisonous aspects and effects of the experimental situation – the ambiguity and uncertainty of prior beliefs and assumptions – so that researchers can capture and represent a phenomenon that might in fact be too complex to be experimentally replicated in the first place.[7]

However, the very cooking practices – the specific ways of imagining data – do not derive only from the methodological thought that first brought the circuits of love into existence and that might indeed reconfigure "evolved structures of practice" as far down as the scanner room. This is to say that the interests of brain imaging's "data monkeys" differ considerably from those of the field's purported stakeholders.

Methodologists typically have a background in either physics, mathematics, computer science, or engineering and often do not regard neuroscience as their final destination. They travel between different fields to gain experience with and cut their teeth on big data from sources such as the Hubble Space Telescope or the Large Hadron Collider at the European Organization for Nuclear Research (CERN) before finding an "application" for their expertise and a home for themselves as researchers. For instance, one of my interviewees, Saïd, a methodologist and postdoctoral researcher employed in a British brain imaging lab, studied engineering in Paris after finishing high school in Morocco without forging close links with his field or developing a distinctive disciplinary identity:

> I was mainly interested in maths at that time and in engineering. I kind of followed the applied maths path, ah, via certain deviations in my head. I think at some point I thought, okay, maths I can study in my spare time, so I could just specialize in civil engineering or aviation or something like that. Something engineery. But I decided not to for various reasons and followed the maths path. So I did applied maths. And when I did applied maths, in the first few weeks they sort of gave us a general introduction into the different sections of the applied maths. So obviously a lot of people want to do finance. But some people came, I can't remember, they weren't from the school, they were from the outside, I can't remember them, but they came and told us about medical imaging in general, not neuroimaging. Never heard of it before, just thought it was a cool application of maths.

Just like Saïd, many methodologists distinguish themselves from "real neuroscientists" by alluding to their command of mathematical skills. Their collaborative promiscuity and the fact that they tend to keep closer ties to "computational people" working in other fields than to "real neuroscientists" at their own institutes are symptoms of a media-technological bias toward statistical techniques and computational thinking, or as Erika Mattila (2005) puts it, "modeling methods" (532). Within this context, the term "big data" does not refer to the size of data sets but to the fact that specific, real-world data might augment and improve the statistical devices that have been designed to inform what in the first instance might look like pure noise.

For example, another interviewee, Steve, the head of analysis in a British brain imaging lab, admitted that only "half of the importance in that work is about the neuroscience relevance of finding networks." Methodologists thus get "sucked into doing bits of neuroscience" (Stan) to the degree that statistical models originating in astronomy and finance are being used in studies of the brain and its workings; however, they typically do not "convert" to the field given that it might provide them with only a temporary home.[8] Johan explained, "Yeah, from my perspective, my contribution to the field will be more in terms of, like, providing new tools, finding new mathematical ways of solving the problem of doing this ... So basically this is kind of an example problem for me to focus on ... What we ideally want in the end is to have interesting neuroscientific results, but we also want to have a method that is innovative and that can be translated and applied to other fields. So, to me, I would have to say the methodological implications are the most important."

In fact, the workings of the brain translate into "cool applications" and provide test cases for statistical devices that are potentially of "general purpose." Many methodologists' work is indeed "data-driven," but their contribution to brain science cannot be reduced to processing as many numbers as possible; it is tied to and inextricably entwined with methodological excursions that cannot be evaluated by the standards of brain imaging and cognitive neuroscience alone. Although illuminating the brain's information processing capacities represents an ultimate goal, brain imaging data first and foremost provide empirical realities for the development of models that might be key, so the argument goes, to solving entirely different and ultimately data-based problems.

Informational labour involves straying beyond the boundaries of the research fields and working with the signal to generate data that "speak" to both the explorative analysis of brain function *and* the goal of refining methods that have been designed to live on the fringes of classic disciplines or the sort of "object-oriented" research that the neurosciences seem to exhibit. For instance, in an opinion article aptly titled "The Need for the Emergence of Mathematical Neuroscience: Beyond Computation and Simulation" (2011),

Gabriel Silva contends that the decisive question for applied mathematics is not what mathematics can do for physics but what physics can do for maths, noting that "a similar argument can be applied to the relationship between mathematics and neuroscience where it is clear that we either are not using the right mathematical tools to understand the brain or such tools have not yet been discovered. The resultant mathematical descriptions should make non-trivial predictions about the system that can then be verified experimentally. This approach takes advantage and has the potential to use the vast amounts of *qualitative* data in neuroscience and to put it in a quantitative context" (2, original emphasis).

To the degree that computer simulation in the mid-twentieth century provided mathematical physicists and statistical devices with what French nuclear physicist Lew Kowarski has called "a new way of life in nuclear science" (quoted in Galison 1996, 139), the brain has emerged as a productive enigma for the sort of data scientists who excel in their mathematical or coding skills and aspire to live within a third culture that links entirely different research objects by means of software suites and statistical devices.

Brain imaging data provide a messy numerical reality that challenges mathematical models to a point where they need to evolve in order to reliably interpret and live up to the ground truth that the workings of the brain provide. Accordingly, noise does not figure as a hindrance to methodological work in brain imaging; instead, it is embraced since it potentially increases the robustness and reach of models. The intriguing novelty of this development is that both the brain and statistical devices come to reconfigure each other within the noisy, datafied realities that brain imaging provides and through the work of researchers who are most "at peace" when they have the opportunity, simply, to "code."

CODING

"I will literally sit there and write down equations and do simulations ... to see how the signal changes and whether the system goes unstable," Rita explained about her favourite part of the job. She described the initial process of going back and forth between the model and its synthetic output as variously the "creative stage" and her "core methods time," a relief from the tasks of a promiscuously collaborating "data monkey." Liberated from the messy empirical realities of volunteers in the scanners, methodologists retreat to their offices and develop ideas or build prototypes within the programming environment of MATLAB or similar digital tools such as SciLab or Octave, which have come to serve the computational implementation of mathematical theorems in a variety of disciplines and research environments.

MATLAB, in particular, has played an important part in configuring the brain for the purposes of data science. As Thomas Haigh (2008) writes, MATLAB's developer, Cleve Moler, had originally envisioned MATLAB as a simple matrix calculator that would relieve his mathematics students from the burden of having to learn programming in Fortran in order to conduct matrix operations on their computers. In conversations with Haigh, Moler explained that he had realized that his students were occupied with the rites of programming to a degree where they lost track of and were unable to concentrate on the mathematics they were supposed to attend to. MATLAB provided a mathematical laboratory that, in its developer's words, might not help with "inventing new mathematics or new mathematical algorithms, but ... [made] mathematical thinking and mathematical techniques accessible to scientists and engineers who might not encounter them otherwise" (91).

A decade ago, MATLAB was still mainly used by practising scientists and engineers in automotive and aerospace firms, but thanks to the gradual addition of specialized toolboxes for signal processing, image processing, bioinformatics, and financial modelling applications, it has meanwhile become an environment that frees brain imaging methodologists from the constraints of mathematics formalism while providing them with the means to "broadcast new ideas" in a readable, tractable, and most significantly, transferable form. Gareth said, "I think the advantage of MATLAB then and now is [that] because it's basically a pseudo code written at a very high level, it's very accessible.[9] So it has the same syntax and readability as a sort of linear algebraic matrix that you would see in a standard textbook ... Furthermore, because it is nice and transparent, it means that people can adjust and modify and nuance the code to their own ends."

Whereas mathematical tools are typically not perfectly malleable and transparent, MATLAB allows modellers to work with theorems that have been successfully employed in different fields, and it provides an environment for modifying parameters or coding new elements without having to rethink models from scratch. One of my interviewees, John, a methodologist in a Swiss brain imaging lab, explained that modelling in MATLAB thus resembles a playful engagement of mathematical formalism, not an abstract reasoning that "could probably be done on paper." As he put it, "it's not really mathematical reasoning. It kind of comes after the mathematical reasoning or before. It's really an intertwined relationship" – a data-driven process of reimagining tried and tested idealizations, omissions, and approximations from foreign fields as elements of novel and potentially disrupting descriptions of the brain and its workings. Distributed in the form of code, packaged for easy implementation by its "users," and equipped with guidelines for good practice in research, MATLAB provides a platform for methodologists' work to be "sucked into ... neuroscience" (Stan) and at the same time draws the brain into the world of applied mathematics and the minds of engineers and data scientists.

Rather than a mere piece of software designed to give experimentalists relief from informational labour, MATLAB hence figures as an interface where the brain first emerges and becomes workable as a cloud of data. In *The Stack* (2015), Benjamin Bratton considers the etymological origins of the "platform," determining that it refers to a plan of action – a scheme or design – that is later conjoined with the plot, which ensnares individuals into an unfolding structure. "Platform" also has a political meaning linked to party policies that suggests the strengthening of that which glues the members of an esoteric circle together and affects those belonging to a complementing exoteric circle of researchers who merely make use of "tools." The platform is neither entirely technical nor characteristically ideological. It merges a more or less elusive and transcending idea into protocols and rules that determine what can be said, but at the same time, a platform does not determine outcomes: "[T]he design logic of platforms is the generative *program*" (44, original emphasis) that unfolds not as a master plan but by turning every user into a contributor to an introverted logic. Computational platforms, if you will, represent a space where methodological thought unfolds.

Particularly via their command of software suites such as MATLAB, which are black-boxed and withdrawn from interventions of the "user" to a certain degree and thus implement standards and conventions of practice, methodologists have the opportunity to act upon the threshold of perception in the brain sciences. Although platforms cannot be overhauled at once, they offer points of leverage for those in the know – twists and tweaks below the surface of an interface that change the basis for the user's perceptions and conceptions from culture to technique. Through a mix of promiscuous collaboration and cooperation characterized by partial autonomy, methodologists thus orchestrate the statistical transformation of the brain by means of their insights into how the differentiation between signal and noise in data can be reconceived. Another meaning of "platform" is "plateau" or "raised level surface" (from Middle French *platte form*): if you are in command of the platform's architecture, as it were, you can always make it sink back down into an ocean of noise.

For instance, the manipulation of the signal in response to empiricisms tested in astrophysics might thus contribute to changing the cognitive threshold in many brain imaging laboratories and the field of cognitive neuroscience in general, which, again, can "make you lose a bit your stability," as Yann put it with regard to his neuroscientific collaborators. This stability is not only that of the collaborator but also that of a suddenly elusive cognitive architecture that derives from imagining data differently. Stan explained,

> So the reason we started doing more advanced modelling was actually because we were given better data and not because we set out to try and achieve the goal of finding temporally independent networks ... MRI physicists came up with basically, yeah, an acceleration. So you could get the functional MRI data much faster, which means we had a lot more time points. And instead of a couple of hundred time points, we now had thousands. And we just sat there for days literally, thinking of different things that one might do to take advantage of that and, eventually after a few full stops going into the wrong directions, realized that we could simply look for, basically use ICA [independent component analysis] with temporal independence, so look for temporally distinct, overlapping networks. And once we realized that basically we were doing temporal ICA, of course we realized that that's no genius because people have been using temporal ICA in other domains. But that meant that we could for the first time start looking at the temporal dynamics of these resting state networks.

Although both developing statistical techniques employed to speed up acquisition and increasing the quality of the signal through compression might, at face value, appear to be mere computational support tasks, they potentially turn into valuable or even game changing informational labour if they allow for a novel empiricism to emerge. In the example above, an increase of noise in the data eventually initiated the use of techniques for modelling temporal dynamics of interacting networks instead of merely mapping their components. As Stan articulated, this reorientation was not due to a sudden strike of genius but to a meeting of brain imaging signals and statistical techniques tried and tested in other data-heavy fields.

Images of the brain have always been visualizations of only statistically significant changes in blood flow throughout specific brain structures. The major difference between then and now is the much more open propagation of "data science" methodology in cognitive neuroscience – a methodology, as it were, where "traditional statistical modeling techniques percolate together with a range of other mathematical approaches to characterization and prediction, drawn from fields such as graph theory, complex systems analysis, and mathematical logic" (Lowrie 2017, 4).

To the degree that MATLAB and other programming environments permit integrating dynamics characteristic of brain imaging data into "hypothetical systems" (Knuuttila and Loettgers 2016, 1011), they support the emergence of (novel) esoteric circles whose members do their work, just like the models they build, in between disciplines or specific object-oriented research fields. With increasing noise and elevated complexity, uncertainty takes hold of biological systems, and "data monkeys" increasingly become sought-after, "powerful brokers of information"[10] within a field that they themselves still describe as a foreign environment or simply as a "cool application of maths" (Saïd). It might not be surprising, therefore, that their informational labour introduces a considerable amount of uncertainty not only about the idea of the brain as a biological system but also about the field of cognitive neuroscience.

––––––––––

The following two chapters allude to the younger history of brain imaging and approach changing neuroscientific understandings of the brain's functional organization through past controversies over the use of imaging technologies and statistical devices. These controversies were intimately linked to the informational labour of brain imaging methodologists, who suddenly figured as stakeholders rather than as mere "data monkeys" and consequently improved their position within the field of cognitive neuroscience. Hence, to convey how informational labour may have repercussions for "real neuroscience," I now go backward from data analysis to the experiments where data are first generated.

To start with, I discuss how the statistical malpractice of "double-dipping," "cherry-picking," or "p-hacking" has contributed to "false positives" in social and cognitive neuroscience, thus exposing brain imaging as well as the neuroscientific investigation of sociality as a highly technical process at heart. Indeed, many neuroscientists were reluctant to take part in an open debate in the first place since the statistical problems tackled in the so-called voodoo controversy were not readily accessible for everyone involved. Even Edward Vul, the whistleblower himself, admitted that he initially merely understood that "there was something fishy going on" (interviewed in Lehrer 2009) with high correlations between behavioural measures and brain activations. The easiest solution to the problem thus seemed to be to sit it out and wait till methodologists had cleared the air. Daniel Margulies (2011) reports a telling example of how the anxieties that had arisen about the field were typically erased in the aftermath of the debate. He had received a text message from a colleague in New York that read, "Went out for drinks with the stats department. Comfortable now that voodoo correlations argument is bullshit" (278).

It might be surprising that neuroscientists confronted with statistical techniques for data analysis on a daily basis needed to go out for drinks with the stats department in order to assess the validity and significance of the voodoo allegations. As I have tried to show throughout this chapter, however, the internal "stats departments" of neuroscience institutes often figure as a strangely isolated place. The decision about what "counts" in sets of brain imaging data typically falls within the remit of methodologists who develop and implement the statistical devices that turn pure noise into insights about the workings of the brain. Easy access to methodological advice implies that many researchers do not need to grapple with, and are often not entirely familiar with, methodological details and statistical intricacies.

As shown in chapter 3, the significance of an at least rudimentary understanding of computational modelling or the ability to "think in MATLAB," as Gareth put it, imposes constraints on the day-to-day interaction among methodologists and experimenters, or "real neu-roscientists," and thus on research in cognitive or social neuroscience in general. The voodoo controversy put many neuroscientists into a position where they could not simply dismiss accusations levelled against leading researchers in the field of social neuroscience since the statistical techniques in question had apparently been routinely applied throughout the field. A pre-print for a scientific article, writ-ten by psychologists, incited a short-lived yet far-reaching debate that first unsettled the field of social neuroscience and subsequently disrupted cognitive neuroscience in general. In an attempt to isolate "the social" within the brain, researchers had "cooked" their data to a degree where all flavours had evaporated – a sudden loss of infor-mation that cast doubt on the epistemologies of an entire field and, subsequently, on commonly accepted notions of brain architecture or, in other words, our "image" of the human brain.

3

False Positives

On a banner that hovers above his latest blog posts, Will Gervais identifies as an evolutionary and cultural psychologist, currently employed by the University of Kentucky. His name, written in bold blue letters, is flanked by two black and white images of churches that provide a fitting atmosphere to a blog post about Gervais's (2017) "methodological awakening" and related afterthoughts. To cut a long story short, I offer a line from his post: "FFFFFFFFF-FUUUUUUUUUUUUUCCCCCCCCCCKKKKKKKKKKKK!!!!" And here is another one: "FFFFUUUUUCCCCCKKKKK YOOOOOOOOOUUUUUUU FORMER WILL!!!"

Julia Rohrer, Tal Yarkoni, and Christopher Chabris call such methodological awakenings a loss of confidence in the withering findings of individual experiments or studies. In order to give researchers the opportunity to publicly confess doubts over the results of studies that they themselves conducted, the three psychologists created a website that provides a Loss-of-Confidence Project submission form and definitions of what might count as a proper loss of confidence.[1] Submitting does not automatically imply that the respective paper will be retracted; after all, the project's most important goal is to show how the results of experiments conducted in good faith can "wither" over time, particularly when specific methodologies fall out of fashion within a field.

This is exactly what happened to Will Gervais and, if we believe the initiators of the Loss-of-Confidence Project, among others, to more or less the entire psychological research community. Rohrer, Yarkoni, and Chabris's project, after all, was born in reaction to diagnoses of a replicability crisis in psychology, which has become ever more apparent during the past decade. Epidemiologist John Ioannidis (2005) had already published a now infamous paper where he explained why most published research findings are false according to a statistical logic. A follow-up paper by Ioannidis (2012) then proclaimed that "science is not necessarily self-correcting," and this paper was complemented by two others written by Daniele Fanelli (2010, 2012), who diagnosed a clear bias toward reporting only spectacular and novel findings in most disciplines. He stated that negative results (where the hypothesis cannot be confirmed through a study) or attempts to replicate earlier findings are basically absent from leading journals in many disciplines,[2] noting that psychology apparently makes for the worst offender in this regard.

That our contemporary knowledge industry privileges odd and spectacular findings might not be surprising and is generally not (too) problematic; the replicability crisis in psychology, however, became so pressing around 2012 because researchers within the field started to realize that the results of many published studies could not be replicated (Yong 2012). Joseph Simmons, Leif Nelson, and Uri Simonsohn's paper "False-Positive Psychology" (2011) consequently attributed issues with replicability to the surprisingly prevalent practice of "p-hacking" (see also Simonsohn, Nelson, and Simmons 2014).[3] Reflecting on the success of their initial paper, the authors write, "We knew many researchers – including ourselves – who readily admitted to dropping dependent variables, conditions, or participants so as to achieve significance [see figure 3.1]. Everyone knew it was wrong, but they thought it was wrong the way it's wrong to jaywalk. We decided to write 'False-Positive Psychology' when simulations revealed it was wrong the way it's wrong to rob a bank" (Simmons, Nelson, and Simonsohn 2018, 255).

Similar concerns about "data dredging," "data snooping," "data fishing," or "post-hoc data-torturing practices" (Yong 2012) had emerged within the social neuroscience community some years before, and some of the protagonists of the short-lived voodoo controversy have indeed been involved in psychology's replicability crisis as well.[4] In this chapter, I show what has changed since an article by Edward Vul and colleagues (2009a), considered "a bombshell of a paper" (Bell 2008), hit the grounds of the neurosciences. The latter four sections of the chapter provide a detailed analysis of how the voodoo controversy unfolded and was smoothly resolved only six months later, concluding with a comparison of social neuroscience's

Figure 3.1
xkcd comic that
mocks "data
dredging."

fall from grace and the replicability crisis in psychology. I put an emphasis on the quick and smooth resolution of the voodoo controversy, which exposed how the epistemically productive division of labour characteristic of brain imaging research (portrayed in chapter 2) allowed the problem of puzzlingly high correlations in social neuroscience to be reduced to a mere technicality. Relief was fittingly provided, as I show in the next section, by the surprisingly active brain of an Atlantic salmon, which was not alive when it was put into the scanner. A unifying mascot, as it were, that nevertheless cast a shadow of doubt on the sparsely activated brain and on the functionally segregated human brain, and that eventually destabilized an entire imaging paradigm and the epistemologies of brain mapping in general.

THE MARVELLOUS SUCCESS
OF A DEAD ATLANTIC SALMON

In June 2009 a poster that showed an image of the allegedly active brain of an Atlantic salmon was put up at the Annual Meeting of the Organization for Human Brain Mapping in San Francisco (see figure 3.2). A sizable number of people stopped by, and the poster was honourably mentioned by leading neuroscientist Rainer Goebel as one of his favourites during a wrap-up of the meeting. The poster had piqued enormous interest not because of the revolutionary results it supposedly communicated but thanks to the fact that the salmon was already dead when the experiment was conducted: according to the poster, the salmon weighed 3.8 pounds, was 18 inches long, and was not alive at the time of scanning.

Scanning inanimate objects, or "phantoms," to test new MRI protocols and configure the scanner for a series of experiments is common practice in brain imaging, and despite the fact that using dead animals instead of objects is rather rare, such a scan would normally not be worth a poster at a brain imaging conference. In this case, however, the researchers had not only scanned a dead salmon but had also turned it into the subject of a typical social neuroscience experiment. The authors of the poster, Craig Bennett, Abigail Baird, Michael Miller, and George Wolford (2009), subjected the salmon to a mentalizing task in order to measure activations in the salmon's brain and determine the neuronal correlates of social cognition: "The task administered to the salmon involved completing

Neural correlates of interspecies perspective taking in the post-mortem Atlantic Salmon: An argument for multiple comparisons correction

Craig M. Bennett[1], Abigail A. Baird[2], Michael B. Miller[1], and George L. Wolford[3]

[1] Psychology Department, University of California Santa Barbara, Santa Barbara, CA; [2] Department of Psychology, Vassar College, Poughkeepsie, NY; [3] Department of Psychological & Brain Sciences, Dartmouth College, Hanover, NH

INTRODUCTION

With the extreme dimensionality of functional neuroimaging data comes extreme risk for false positives. Across the 130,000 voxels in a typical fMRI volume the probability of a false positive is almost certain. Correction for multiple comparisons should be completed with these datasets, but is often ignored by investigators. To illustrate the magnitude of the problem we carried out a real experiment that demonstrates the danger of not correcting for chance properly.

METHODS

Subject. One mature Atlantic Salmon (Salmo salar) participated in the fMRI study. The salmon was approximately 18 inches long, weighed 3.8 lbs, and was not alive at the time of scanning.

Task. The task administered to the salmon involved completing an open-ended mentalizing task. The salmon was shown a series of photographs depicting human individuals in social situations with a specified emotional valence. The salmon was asked to determine what emotion the individual in the photo must have been experiencing.

Design. Stimuli were presented in a block design with each photo presented for 10 seconds followed by 12 seconds of rest. A total of 15 photos were displayed. Total scan time was 5.5 minutes.

Preprocessing. Image processing was completed using SPM2. Preprocessing steps for the functional imaging data included a 6-parameter rigid-body affine realignment of the fMRI timeseries, coregistration of the data to a T_1-weighted anatomical image, and 8 mm full-width at half-maximum (FWHM) Gaussian smoothing.

Analysis. Voxelwise statistics on the salmon data were calculated through an ordinary least-squares estimation of the general linear model (GLM). Predictors of the hemodynamic response were modeled by a boxcar function convolved with a canonical hemodynamic response. A temporal high pass filter of 128 seconds was include to account for low frequency drift. No autocorrelation correction was applied.

Voxel Selection. Two methods were used for the correction of multiple comparisons in the fMRI results. The first method controlled the overall false discovery rate (FDR) and was based on a method defined by Benjamini and Hochberg (1995). The second method controlled the overall familywise error rate (FWER) through the use of Gaussian random field theory. This was done using algorithms originally devised by Friston et al. (1994).

DISCUSSION

Can we conclude from this data that the salmon is engaging in the perspective-taking task? Certainly not. What we can determine is that random noise in the EPI timeseries may yield spurious results if multiple comparisons are not controlled for. Adaptive methods for controlling the FDR and FWER are excellent options and are widely available in all major fMRI analysis packages. We argue that relying on standard statistical thresholds (p < 0.001) and low minimum cluster sizes (k > 8) is an ineffective control for multiple comparisons. We further argue that the vast majority of fMRI studies should be utilizing multiple comparisons correction as standard practice in the computation of their statistics.

REFERENCES

Benjamini Y and Hochberg Y (1995). Controlling the false discovery rate: a practical and powerful approach to multiple testing. *Journal of the Royal Statistical Society: Series B*, 57:289-300.

Friston KJ, Worsley KJ, Frackowiak RSJ, Mazziota JC, and Evans AC. (1994). Assessing the significance of focal activations using their spatial extent. *Human Brain Mapping*, 1:214-220.

GLM RESULTS

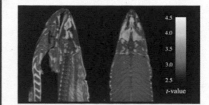

A *t*-contrast was used to test for regions with significant BOLD signal change during the photo condition compared to rest. The parameters for this comparison were $t(131) > 3.15$, p(uncorrected) < 0.001, 3 voxel extent threshold.

Several active voxels were discovered in a cluster located within the salmon's brain cavity (Figure 1, see above). The size of this cluster was 81 mm³ with a cluster-level significance of p = 0.001. Due to the coarse resolution of the echo-planar image acquisition and the relatively small size of the salmon brain further discrimination between brain regions could not be completed. Out of a search volume of 8064 voxels a total of 16 voxels were significant.

Identical *t*-contrasts controlling the false discovery rate (FDR) and familywise error rate (FWER) were completed. These contrasts indicated no active voxels, even at relaxed statistical thresholds (p = 0.25).

VOXELWISE VARIABILITY

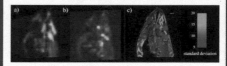

To examine the spatial configuration of false positives we completed a variability analysis of the fMRI timeseries. On a voxel-by-voxel basis we calculated the standard deviation of signal values across all 140 volumes.

We observed clustering of highly variable voxels into groups near areas of high voxel signal intensity. Figure 2a shows the mean EPI image for all 140 image volumes. Figure 2b shows the standard deviation values of each voxel. Figure 2c shows thresholded standard deviation values overlaid onto a high-resolution T_1-weighted image.

To investigate this effect in greater detail we conducted a Pearson correlation to examine the relationship between the signal in a voxel and its variability. There was a significant positive correlation between the mean voxel value and its variability over time (r = 0.54, p < 0.001). A scatterplot of mean voxel signal intensity against voxel standard deviation is presented to the right.

Figure 3.2 The dead salmon poster: "Neural correlates of interspecies perspective taking in the post-mortem Atlantic Salmon: An argument for multiple comparisons correction" (Bennett et al. 2009).

an open-ended mentalizing task. The salmon was shown a series of photographs depicting individuals in social situations with a specified emotional valence. The salmon was asked to determine what emotion the individual in the photo must have been experiencing. Stimuli were presented in a block design" (Bennett, Miller, and Wolford 2010, 2).

Quite surprisingly, the researchers found the dead salmon's brain to be activated during the task, yet they reported – making use of the typical neuroscientific jargon – that "due to the course resolution of the echo-planar image acquisition and the relatively small size of the salmon brain further discrimination between brain regions could not be completed" (Bennett, Miller, and Wolford 2010, 2). The real clue, however, was hidden in the method section: Bennett and colleagues had not corrected for multiple comparisons.

The multiple comparisons problem is a well-known issue with statistical analyses and describes an increase in the number of false correlations – also called false positives – as a result of considering a set of statistical inferences simultaneously. The greater the number of statistical tests performed, the higher the chance that correlations will appear significant even though they are not. A drug, for instance, might erroneously appear to have an effect on a disease when tested on a number of symptoms without increasing the threshold of individual tests, the reason being that statistics are always just probabilities and approximations.

A typical statistical threshold, $p \leq 0.05$, implies that roughly 5 out of 1,000 tests are expected to show a false effect, such as that caused by noisy fluctuations in neuronal activity. Since a typical MRI scan contains roughly 1 million voxels that are each 1 millimetre in size, 5,000 voxels might be incorrectly classified as significantly activated during an experimental task if the data are not corrected for multiple comparisons. If you are unlucky, said false positives can be grouped together within a specific functional brain region – or within the brain of a dead salmon. Bennett had expected to find false positives when reanalyzing the data, yet he was not prepared to find three significant voxels arranged together in the salmon's brain. "If they would have been anywhere else the salmon would have been just a curious anecdote, but now we had a *story*" (Bennett, Miller, and Wolford 2010, 2, original emphasis). Omitting correction for multiple comparisons can make it seem as though one has brought the brain of a dead Atlantic salmon back to life.

Bennett and colleagues' (2009) poster was a huge success and raised eyebrows not only in the neuroscience community. Critical neuroscientist Daniel Margulies (2011) called the salmon poster "the most globally appreciated prank to ever make use of an fMRI scanner" (282) and optimistically assumed that it might mark the beginning of the "decade of ironic neuroscience" (283). Without any doubt, the poster was a very intelligent way of pointing out the pitfalls of data analysis methods in brain imaging. Its success, however, at least partly derives from the time of publishing and the psychological effect it therefore had on all those who saw the success story of brain imaging come to an abrupt end.

Indeed, the salmon brain data were anything but fresh at the time of publishing. Already in 2005, during his first year of graduate school at Dartmouth College, Bennett and his colleague Abigail Baird had subjected a Cornish game hen and an Atlantic salmon to a brain scan in a challenge "to scan the most curious objects [to be found] at the local grocery store" (Bennett 2009). The salmon data had subsequently been stashed away on Bennett's hard drive and was unfrozen when the young researcher discussed the multiple comparisons problem with his co-advisor George Wolford.

In January 2009, just after another "bombshell of a paper" (Bell 2008) that tackled similar problems was dropped onto the neuroscience community, Wolford convinced his initially reluctant doctoral student to put the fish out. Due to the publication of a pre-print article written for a special issue of *Perspectives on Psychological Science* by a young researcher of cognitive science, Edward Vul, and his colleagues (2008) at the Massachusetts Institute of Technology, intense discussion about the methodological repertoire of social neuroscience had set in. In the article, titled "Voodoo Correlations in Social Neuroscience," leading social neuroscientists were accused of forgoing necessary statistical techniques for error correction in the analysis of brain imaging data. For the field of social neuroscience, the article proved to be extraordinarily dangerous since a nuanced debate about the substance of its criticism seemed impossible after it had been published online just before Christmas. The sensationalist title alarmed the scientific public and made the press jump onto the bandwagon with a slight but noticeable delay.

"JUST TYPE 'VOODOO CORRELATIONS' INTO GOOGLE"

The first time I heard about the "voodoo paper" was during my interview with Dejan. Discussing the pros and cons of various approaches to visualizing brain imaging data, Dejan suggested that visualizations by their nature are never "seductive" (Weisberg et al. 2008; see chapter 1) or misleading. Handled with care and supplemented with in-depth information for the most significant results, they enrich the research process and reveal new insights into brain function. The exceptions, of course, are cases where the data are flawed – which is where the "voodoo paper" came into play. Dejan told me, "There is actually a paper which has originally a title like 'Voodoo Correlations in Social Neuroscience,' you know that one? The author is Vul, I think. So you just type 'voodoo correlations' into Google. Well, it got another title when it was published, but there is still a webpage with this title. So the guy pinpoints that there were a lot of papers in this field where they did first statistical tests and then they extracted a measure, and then they did a second statistical test and then they see how beautiful the results of the second statistical test are – which is wrong."

What the then early career psychologist Edward Vul and his co-authors describe in their paper – eventually titled "Puzzlingly High Correlations in fMRI Studies of Emotion, Personality, and Social Cognition" (2009) – is the methodologically problematic practice of "double dipping" (Kriegeskorte et al. 2009, 535), which is selecting statistically significant correlations and performing a second statistical test only on correlations that have been identified as significant in the first place. In general, conducting two statistical tests in a row – selection and selective analysis – does not necessarily represent methodological malpractice; under certain circumstances, however, the results will be unduly inflated. In extreme cases – which are quite numerous in many fields and disciplines, if we believe the authors – double dipping produces false positives and turns erroneous claims into significant findings.

Devaluing scientific techniques and methodologies has ever been a successful and popular strategy to gain epistemic ground in heterogeneous fields. Opponents in a battle for epistemic authority are used to destabilizing each other's accounts by unmasking purportedly flawed experimental designs, misguided interpretations, and the use of mere rhetorical figures as an alleged proof, while projecting their very own truth claims as objective facts.

Controversies make sciences look vulnerable and dysfunctional, yet they are epistemically productive in that they facilitate the rise and demise of paradigms and the restructuring of entire research fields (Brante, Fuller, and Lynch 1993; Collins 1981; Collins and Pinch 1993; Dascal 1998, Engelhardt and Caplan 1987; Latour 1987; Machamer, Pera, and Baltas 2000; Myers 1990; Thomas 2009). "In the course of disputes, the special interests, vital concerns, and hidden assumptions of various actors are clearly revealed" (Nelkin 1992, vii) and hence subject to open discussion and rigorous scrutiny. Typically, such methodological controversies unfold over a longer period of time and come to delineate different strands and approaches to a common research object until one party wraps it up and provides closure to those in disarray.

The voodoo controversy was different. Launching a pointed attack on leading figures in the field by first publishing their paper online under the title "Voodoo Correlations in Social Neuroscience" (2008), Vul and colleagues made it look as though social neuroscience in general was to blame, thus putting an enormous amount of pressure on the brain imaging community as well. Many researchers alluded to the "voodoo paper" when I spoke with them about the pros and cons of contemporary neuroscience and brain imaging methodologies. Typically, my interlocutors sided either with Vul and his co-authors or with those who did not hesitate to respond but instead launched counterattacks through various channels around the turn of 2008. An outstanding source of arguments for demarcation, the voodoo controversy continued to dominate the discursive landscape in brain imaging institutes after it had been resolved. Mentioning voodoo correlations in conversations oftentimes changed the atmosphere: people became more engaged in discussions, and some showed a deep and almost emotional involvement with the issue.

What had caused the stir, apart from a sudden surge in attention for social neuroscience and brain imaging in general, was in fact a widespread uncertainty about the legitimacy of grave accusations levelled against leading figures in the field. Many neuroscientists had been caught on the wrong foot, as widely and routinely used techniques of statistical testing were debunked on the grounds that they inflated the scale of the effects studied in brain imaging experiments.

When a "bombshell of a paper" (Bell 2008) was dropped on the research community shortly before Christmas, it made many neuroscientists fear for their careers.

A BOMBSHELL DROPS

For researchers in social neuroscience, the year 2008 ended with a big bang. Science blogs all over the Internet reported that an article, which had been accepted for publication in *Perspectives on Psychological Science* and was therefore made available online for comment, accused many researchers in the field of using improper statistical techniques in the analysis of functional brain imaging data. The article's authors had reached their conclusion by analyzing correlations between experimental tasks and activations in the brains of volunteers reported in a selection of social neuroscience papers. If the article had not carried the title "Voodoo Correlations in Social Neuroscience" (Vul et al. 2008), these accusations might have made for an also-ran and gone unnoticed by the wider public. The decidedly combative title, however, provoked sensationalist headlines such as "Scan Scandal Hits Social Neuroscience" (Neurocritic 2008) and "Vul on fMRI Abuse in the Cognitive Neuroscience of Social Interaction" (joneilortiz 2008) throughout the blogosphere. It seemed as though many science bloggers had been waiting for a chance to strike against neuroscientists' apparently unconditional faith in functional brain imaging data.[5]

The successful yet infant branch of "social neuroscience" in particular was looked at suspiciously by many authors of science blogs and made an easy target for criticism.[6] What had before been missing, however, was proof – facts that would reveal that analysis methods employed to identify statistically significant brain activations within an ocean of noisy data had contributed to projecting the extremely complex and capacious concept of sociality onto a surprisingly well-delineated social brain (Neuroskeptic 2009). And suddenly, a young researcher from MIT and his colleagues had provided the necessary ammunition by declaring that social neuroscience papers were, and would continue to be, plagued by "voodoo correlations."

In an interview with Jonah Lehrer (2009) for *Scientific American*, lead author Edward Vul retrospectively shed some light on what had first prompted him, as Lehrer put it, to take "a critical look at fMRI papers in social neuroscience":

Some four years ago [University of California at San Diego neuroscientist] Hal Pashler and I saw a talk in which a very high correlation was reported between brain activity and the speed with which someone walked out of the room after the study. Given what we knew about fMRI and the factors that determine how quickly we tend to walk in general, it seemed unbelievable to us that activity in this specific brain area could account for so much of the variance in walking speed. Especially so, because the fMRI activity was measured some two hours before the walking happened. So either activity in this area directly controlled motor action with a delay of two hours – something we found hard to believe – or there was something fishy going on.

Such "fishy" correlations did indeed prove to be not entirely uncommon throughout the field of social neuroscience.[7] After Nancy Kanwisher, whose lab at MIT Vul had joined in the meantime, brought to his attention a further paper sporting mysteriously high correlations between emotional states and brain activations in brain imaging studies, the young researcher felt prompted to analyze the methodological integrity of a representative number of social neuroscience papers. Based on information collected by means of questionnaires, Vul and colleagues (2009a) found that most of the analyzed papers' authors had used statistical techniques improperly, consequently inflating their results. Were social neuroscientists and brain imagers in general guilty of summoning highly significant correlations between sociality and what happened in the brains of experimental subjects during behavioural experiments?

Apart from the aforementioned issues with multiple comparisons, blacklisted papers apparently suffered from nonindependent selections of data chunks – also known as "double dipping." Whereas taking a data set, picking the highest numbers, and performing a correlation analysis are steps in a perfectly legitimate procedure, reporting the high correlation for the whole data set will result in highly inflated correlations. "With enough voxels, such a biased analysis is guaranteed to produce high correlations even if none are truly present. Moreover, this analysis will produce visually pleasing scattergrams that will provide (quite meaningless) reassurance to the viewer that s/he is looking at a result that is solid" (Vul et al. 2009a, 279).

Indeed, Vul and colleagues (2009a) found a remarkable 54 per cent of the scrutinized papers guilty of nonindependent selections and stated that "the problem is exacerbated in the case of the 38% of our respondents who reported the correlation of the *peak voxel* (the voxel with the highest observed correlation) rather than the average of all voxels in the brain" (281).

By addressing specific papers and their authors, Vul and colleagues precluded that any of the accused researchers could ignore the debate or quietly correct their mistakes. Confronted with this "sniper-like tactic," the authors of blacklisted papers were faced with three options: "(1) admitting the wrong and re-analyzing their data; (2) pleading innocence and demonstrating the error in the 'Voodoo Correlations' critique; or (3) hoping that nobody had noticed" (Margulies 2011, 275).

LIMITING THE DAMAGE

Although the chances that "Voodoo Correlations in Social Neuroscience" would go unnoticed were already very low in December 2008, the first mainstream article, written by Sharon Begley (2009) for *Newsweek*, cast off any doubts about its impact on the wider scientific public. An article in *Nature* followed (Abbott 2009). On 13 January 2009, the first due rebuttal of the allegations appeared, with others quickly being published across the blogosphere. A group of social neuroscientists took the initiative and openly denied the validity of most of Vul's criticism. In that article, entitled "Response to 'Voodoo Correlations in Social Neuroscience' by Vul et al. – Summary Information for the Press" (2009), Mbemba Jabbi and colleagues list eight counterarguments and explain that the field has much more to offer than statistical analyses: "[T]he paper by Vul et al. uses statistical arguments that are partially flawed and misleadingly implies that social neuroscience studies rest entirely on the sort of brain-behaviour correlations that are criticised" (1). Counter to the critics' argument, "a key question is often not *how strongly* two measures were correlated, but *whether* and *where in the brain* such correlations may exist" (3, original emphasis).

Vul and colleagues (2009b) immediately published an online reply to the rebuttal and refuted the authors' arguments: "The authors suggest that the magnitude of the correlations really doesn't matter, implying that providing accurate measurements of correlations is not all that vital. We addressed this issue in our paper, and explained why effect size measures are of great relevance to both the practical and theoretical importance of findings in any field. Whether or not the authors themselves care about the magnitude of the correlations, their procedures for producing these correlation estimates produce inflated numbers. The scientific literature should, where possible, be free of such erroneous measurements." Jabbi and colleagues continuously updated their rebuttal without tackling Vul's response, thus adding to the impression that the action taken in defence of the field was everything but well prepared. Margulies (2011), for example, judged the rebuttal to represent "the most classically 'improper' scientific proceeding during the month of January" (278).

Although the initial responses challenged the accusations of Vul and colleagues rather unsuccessfully, the arrival of the first invited comments at the end of January changed the situation entirely. The editor of *Perspectives on Psychological Science*, Ed Diener, had invited cognitive scientists, statisticians, neuroscientists, and psychologists familiar with the matter to comment on Vul and colleagues' paper. Matthew Lieberman, Elliot Berkman, and Tor Wager directly referred to the voodoo accusations, titling their response "Correlations in Social Neuroscience Aren't Voodoo" (2009). Theirs was by far the most critical comment, and the fact that Lieberman had co-authored some of the articles blacklisted by Vul and colleagues surely played a role in the fierceness of their rebuttal.

Admitting that the inferential two-step procedure described by Vul and colleagues represents bad scientific practice, the authors surprisingly deny that anyone in the field actually makes use of it and pillory the authors of "Voodoo Correlations in Social Neuroscience" for their purportedly anchorless accusations:

> The word *voodoo*, as applied to science, carries a strong and specific connotation of fraudulence, as popularized by Robert Park's (2000) book, *Voodoo Science: The Road from Foolishness to Fraud*. Though the title was subsequently changed to remove the word *voodoo*, the substance of the article and its connotations are unchanged: It is a pointed attack on social neuroscience. Much of the article's prepublication impact was due to its aggressive tone, which is nearly unprecedented in the scientific literature and made it easy for the article to spread virally in the news. Thus, we felt it important to respond to both to the tone and to the substantive arguments (Lieberman, Berkman, and Wager 2009, 299).

According to the authors, it is ironic that Vul and colleagues, who criticize shaky statistical reasoning in social neuroscience, achieved their popularity based on problematic claims about the process of statistical inference. Descriptive statistics are known to be inflated but not to the reported extent. Further, Lieberman, Berkman, and Wager (2009) rebut the assertion that correlations above 0.74 in social neuroscience are impossibly high; even attending to Vul and colleagues' criteria, studies reporting reliabilities above 0.90 both for brain regions in fMRI (Fernandez et al. 2003; Poldrack and Mumford 2009) and for social distress measures (Oaten et al. 2008; van Beest and Williams 2006) suggest that theoretical upper limits can be well above the suggested 0.74.

Whereas some commenters (partly) accepted the "voodoo" critique, most found the findings less puzzling than suggested, the arguments overstated, and the discussion outdated (Nichols and Poline 2009). Many researchers ironically adopted a social constructivist perspective on scientific practice and science communication while sticking to the belief that data contain a solid and accessible truth. Psychologist Lisa Feldman Barrett (2009) neatly summarizes the general sentiment within the field by pointing to similar issues in the early days of psychology, claiming that "[i]n 70 years, when someone writes the history of how measurements of the brain eventually translated into knowledge about the mind, psychologists and neuroscientists will marvel at how far we have come" (318).

In the aftermath of the voodoo controversy, Bennett and colleagues' (2009) salmon played its part extraordinarily well. Whatever the intent of the authors when they presented their poster at the Annual Meeting of the Organization for Human Brain Mapping, what appeared to be another critique of brain imaging methodology in fact helped to put researchers in the field at ease. Although Bennett and colleagues' poster was conceived as a reminder to the community that "analysis is often complex and always based on assumption" (Kriegeskorte et al. 2009, 535), it eventually turned into an alternative means of qualifying Vul and colleagues' critique, dissolving it within an ironic take on the matter. Margulies (2011) has therefore called Bennett's subject the "salmon of doubt" – a unifying mascot that finally resolved the "fishy" business of inflated correlations after half a year of disquietude.

Instead of becoming the mascot of a "decade of ironic neuroscience" (Margulies 2011, 283), the salmon was solemnly borne to the grave. In his comment introducing the published version of Vul and colleagues' article in a special issue of *Perspectives on Psychological Science*, editor Ed Diener (2009) endorsed limiting the critique to a statistical or methodological problem, writing that "further discussion of the issues should now take place in journals that are focused on imaging and neuroscience" (273). The controversy was resolved, as it were, by shifting the problem (Pinch and Bijker 1984) away from the summoning of a social brain through the "hypothetical machines" (Lemov 2010; see chapter 6) of social neuroscience and toward the technicalities of imaging or, more specifically, the uncertainties inherent to – or worse, introduced *by* – statistical reasoning.

Social neuroscience and psychology were off the hook, and voodoo correlations had been relegated to a mere technicality even though the initial allegations seemed to have the potential to upset the very epistemology of brain imaging research. This is not to say that the voodoo controversy did not have repercussions for neuroscience practice; quite the contrary, as the following section shows. The burden of relieving generic uncertainties, however, was put on the backs of the "data monkeys." Symptomatically, Vul and colleagues' (2009a) article, having been renamed and resubmitted, contains nothing but a footnote on the first page that reminds the reader that the paper's title once featured the term "voodoo" (274).

AFTERTHOUGHTS

Edward Vul and Harold Pashler eventually managed to slip the term "voodoo" into the title of their paper "Voodoo and Circularity Errors" (2012) but simultaneously admit that their critique was only partially successful as regards social neuroscience. Nevertheless, the voodoo controversy did have repercussions for methodological practice in brain imaging, which can now be observed as they ripple through the field of psychology. In "A Short (Personal) History of Revolution 2.0" (2015), Barbara Spellman, who succeeded Ed Diener as editor of the influential journal *Perspectives on Psychological Science*, describes psychology's replicability crisis as a revolution analogous to political revolution, arguing that it will most likely turn into a continuing evolution of psychological science.

Spellman's (2015) optimism derives from a shift in emphasis since the voodoo controversy from debates about the correctness of statistical techniques to debates about a science 2.0, or "open science" (887), which is based on data sharing, methodological disclosure, and embracing uncertainties. This is to say that the replicability crisis has, at times, been not so different from the voodoo controversy: whereas some sensed the end of a paradigm, others saw no more than a tiny statistical issue. Spellman refers to said disparate understandings of methodological crisis by quoting Leonard Cohen at the beginning of her article: "There is a war between the ones who say there is a war and the ones who say there isn't" (886).

Indeed, just as was the case with the voodoo controversy, the very existence of a replicability crisis in psychology has been denied by an array of researchers from the field. In reaction to the concerted efforts of the Open Science Collaboration (OSC) (2015) to replicate as many psychological studies as possible, Daniel Gilbert and colleagues (2016) at Harvard University compiled a technical comment in which, on statistical grounds, they refute the claim that the replicability of psychological studies is surprisingly low. They argue that the OSC paper seriously underestimates the reproducibility of psychological science since the researchers did not account for expected failures (i.e., the aforementioned possibility of false positives or negatives), relied on poorly reproduced studies with small sample sizes, and fell victim to a bias that resulted from their failure to randomly select the results they tried to replicate.

In short, Gilbert and colleagues (2016) attempted to hoist the OSC with its own petard, provocatively stating that "metascience is not exempt from the rules of science" (1037). The OSC, in turn, responded that it considers the reanalysis of the Harvard group itself to be flawed by statistical misconceptions and by the selective examination of data (Anderson et al. 2016). According to the OSC, however, the question must not be reduced to a battle of analysis experts. Instead, it argues that failures to replicate the results of psychological studies should be published and should incite new attempts to generate evidence, to strengthen confidence in published results, and above all, to define commonly shared standards for replication. Spellman (2015) is therefore confident that psychology will come to a sensible middle ground after the revolution through concerted efforts with the social and life sciences to solve the issues introduced through changing technologies, changing demographics of researchers, limited resources, and misaligned incentives.

Although psychology's replicability crisis might have been quite similar to the voodoo controversy in many instances – for it involved a back and forth between two camps who argued about the determinants of good scientific practice by recourse to statistical techniques – it differed from its predecessor in one significant aspect: whereas psychologists started initiatives to replicate studies and to publish replication attempts (e.g., the OSC and PsychFileDrawer)[8] or to publicly confess doubt about the results of their own studies (e.g., the Loss-of-Confidence Project)[9] and went on to develop requirements for authors and guidelines for reviewers (e.g., Simmons, Nelson, and Simonsohn 2011), social neuroscientists remained remarkably inactive during the early months of 2009.

In her study of "the social" in social neuroscience, sociologist Svenja Matusall (2012) reports that the belief in the validity of brain imaging data was indeed barely affected by the voodoo controversy, for the proponents of social neuroscience methodology had done their best to cast off any doubts. Admittedly, the complicated nature of establishing statistical significance is demonstrated by the fact that even statisticians are currently debating usefulness of "p-values" and "Bayes factors" for managing uncertainties that derive from experimental measurements, yet this situation also shows how pressing the issue really is.[10] A *Nature* survey on reproducibility, for instance, found that nearly 90 per cent of the 1,576 involved researchers identified a better understanding of statistics as the most significant approach to increasing reproducibility (Baker 2016). As Nelson, Simmons, and Simonsohn (2018) argue, this is the case for the replicability crisis in psychology since researchers "did not learn from experience to increase their sample sizes precisely because their underpowered studies were not failing. P-hacking allowed researchers to think, 'I know that Jacob Cohen[11] keeps saying that we need to increase our sample sizes, but most of my studies work; he must be talking about other people. They should really get their act together'" (515).

Commenting on voodoo correlations in social neuroscience, Lieberman, Berkman, and Wager (2009) anticipated that the debate would be resolved in a similar fashion. Their message to the community of social neuroscientists was loud and clear: cobbler, stick to your last. They said, "There are various ways to balance the concerns of false positive results and sensitivity to true effects, and social neuroscience correlations use widely accepted practices from cognitive neuroscience. These practices will no doubt continue to evolve. In the meantime, we'll keep doing the science of exploring how the brain interacts with the social and emotional worlds we live in" (306).

The developments that exposed false positives first in social neuroscience and subsequently in brain imaging are reminiscent of the trajectory that Paul Edwards (2010) recounts in his book on climate science as a global knowledge infrastructure. Edwards notes that the planetary-scale infrastructure established for sharing global weather data among meteorologists was put under close scrutiny within the context of climate change research. The system had been constructed to serve the purpose of short-term weather forecasts and was found to introduce accidental inhomogeneities into long-term data, which were of minor significance for weather forecasts yet seriously hampered research on climate change. Researchers hence performed "infrastructural inversion" (Bowker 1994, 10) and put the data collection system itself to the test in order to find and correct inaccuracies, effectively taking what had before merely provided tools to support a certain research practice and turning it into an object of research within a new context.

Comparably, brain imaging methodology and thus the methodological thought that had come to undergird experiments with volunteers in cognitive neuroscience were put to the test in the aftermath of the voodoo controversy. Interestingly, Ed Diener (2009) had suggested in his editorial for the special issue of *Perspectives on Psychological Science* that the voodoo controversy might eventually turn into a problem with the so-called BOLD (blood oxygen level–dependent) paradigm: "In addition, there are questions related to what relative blood-oxygen levels actually signify about the mind when they are uncovered." Yet the paper by Edward Vul and colleagues (2009a) had technically not been the incident that encouraged the field to reimagine brain imaging methodology by means of statistical devices. Brain imagers had already embarked on a mission to recover traces of brain activity that would have been discarded by the standards of twentieth-century brain imaging methodology and the BOLD paradigm. It was "ghostly data" bearing links and traces of "something else," as Blackman (2019, 33) phrases it, that put the resting brain back onto the brain map.

In chapter 4, I provide a brief overview of how the BOLD paradigm came into existence. As physiology – or infrastructural measurements of throughput – fell out of fashion, the technologies of positron-emission tomography and fMRI provided the means to infer cognitive activity from the physiological and were consequently wed to the paradigm of cerebral geography. This geography led to a focus on contrasting activations and hence an emphasis on specific statistical techniques that came to suppress significant cognitive activity. Only when the discussion around false positives in brain imaging linked up with the uncertainties that derived from the identification of "false negatives" – or that which had not surpassed brain mapping's threshold of perception – did the paradigm begin to tumble. Only then were imaging technologies reimagined within a novel context. The following interlude provides an entry point to understanding this process, which was kicked off by an unholy alliance between investigators of false positives and false negatives.

INTERLUDE: (THE FAILURE OF) BOLD PROMISES

The voodoo controversy is a pertinent example of how strategies and techniques of uncertainty management are staunchly defended even when their validity is called into serious doubt. As chapter 3 shows, many researchers proved unable to assess the reach of the alleged voodoo correlations or were resistant to do so, for they were too invested in the brain imaging paradigm that practices such as "p-hacking" had come to support. The blogger Neuroskeptic observes that "the paradox of functional MRI is that it's a highly technical method that was developed by physicists and mathematicians, but the people who actually use it are for the most part biologists ... [I]n my experience it's often used like a black box. You run the software and you press the right buttons and a few hours later you get these blobs appearing in the brain. It's not always used with an awareness of what it is" (quoted in Lyon 2017, 1).

Those who had not anticipated the significance of statistical techniques before the voodoo crisis broke loose became ever more aware of the fact that "false positives" might be central to the current paradigms of brain imaging in general. In his own inimitable way, one of my interviewees, Victor, the principal investigator in a British neurology department, made me understand that Edward Vul and colleagues' (2009a) critique of voodoo correlations was considered inconvenient, to say the least.

At the end of what had turned into an exorcist's hour-long tour through the deficits of contemporary cognitive neuroscience, Victor touched upon the voodoo controversy in social neuroscience:

Yeah, yeah. The problem with the "Voodoo Correlations" is that Ed Vul was a kind of junior person, so he hadn't earned his spurs, you know, but he was right. I can tell you a little story about that. Have you spoken to anybody at [name of institute]? Where they are all true believers? The director actually wrote letters when he learned that Ed Vul was looking for tenure in universities. He actually wrote unsolicited letters to those laboratories recommending, and he is an extraordinary powerful man, recommending that they don't employ this guy. That's how crazy it got. Poor bastard, you know, just for being young and mouthy.

Victor's take on the voodoo controversy suggests that what had elicited the fierce rebuttals of social and cognitive neuroscientists was a general sentiment within the field that a young researcher had overstepped his competencies and, crucially, directed the attention away from the core interests of brain science and toward a supposed sidechain of brain imaging research. However, the reactions of the "believers" – who were ready to protect, by all means and "unsolicited letters," the paradigms that had powered brain imaging research for roughly twenty years – bear witness to the graveness of the voodoo correlations critique. Accordingly rare were favourable comments on Vul and colleagues' paper, such as the one where Tal Yarkoni (2009), a psychologist at the University of Texas in Austin, suggests that "it is almost certain that the vast majority of whole-brain correlational analyses (a) identify only a fraction of the effects that really exist in the population, (b) grossly inflate the apparent size of those effects that researchers are lucky enough to detect, and (c) promote a deceptive illusion of highly selective activation" (297).

What is so remarkable about this statement is that the author does not limit his critique to inflated correlations or false positives; indeed, he maintains that this inflation also occurs at the expense of significant activations that have fallen below the perceptive threshold of cognitive neuroscience for decades. Whereas Vul and colleagues' "voodoo correlations" primarily refer to false positives that occur through flawed techniques for hypothesis testing, Yarkoni (2009) maintains that images of contrasting brain activations produced with contemporary fMRI analysis methodology show *in general* no more than a fraction of what should be considered meaningful activity throughout the brain. Said miscategorization of significant brain activity as noise via statistical devices is tightly linked to a brain imaging paradigm that has remained more or less uncontested since researchers started to map the brain's functional regions on the basis of fMRI data. Victor, by contrast, considered the so-called BOLD (blood oxygen level–dependent) response to be a rather speculative measure of cognitive activity and regarded it, to use Ludwik Fleck's (1979) words, as a readiness for "directed perception, with corresponding mental and objective assimilation of what has been so perceived" (99). Victor explained,

> So now we have the field of neuroeconomics,[1] but if I ask you, "What's neuroeconomics added to behavioural economics?" Nothing, absolutely nothing. It's just replacing the word "people" with "brains," and the lie that neuroscience will give you is to say, "Well, if we see the brain areas, we're understanding the mechanism." But there's nothing mechanistic in it. Do you know incidentally that we don't know what the BOLD response is? ... [P]eople do believe that in the end it will be shown to be a function of the spikes in neuronal activity. But there's no reason to believe that at the moment; there's just the belief. Things keep popping into my head, but you know the case of the dead salmon? ... What was really interesting about that was not the result itself – it just showed you that with statistics you can get false positives – but was the reaction of the neuroscience community. And it really was cultish; there were threats on the Internet. I don't know if you ever saw any of them, people saying, "Oh people always take down something that is successful" [or] "What these people are doing is throwing out the baby with the bath water." But there was no single response like "That's kind of interesting."

Although the BOLD response does in general merely denote a decrease of the magnetic properties of haemoglobin in response to an increase of oxygen in cerebral blood flow, it had become widely accepted as an indicator for cognitive activity represented by co-activations of distinct neuronal populations in response to external stimuli. The BOLD paradigm was hence nearly synonymous with brain imaging in the tradition of brain mapping, for it provided measurements of changes in blood flow with a good spatial resolution. Anyone interested in delineating functional brain regions relied on the BOLD promises of task-based fMRI and tried to ignore the fact that, for instance, the relation between increased blood flow and firing neurons was (and still is) anything but clear; after all, an array of statistical techniques had been developed to make up for the deficits of the BOLD response as a measurement of brain activity. Software tools contributed their share to cementing the significance of the BOLD response for brain mapping since standard configurations allowed "new users [to] inadvertently load a host of assumptions at the click of a button" (Carlson et al. 2018, 95), thereby turning incredibly "noisy" neuronal chatter into contrasting brain activations.[2] As Victor observed, critiquing the BOLD response, or merely the statistical techniques that support it, was consequently conceived as challenging the experimental paradigm of cognitive neuroscience in general.

Therefore, the discovery of false positives in brain imaging did amount to a loss of contrast in brain imaging and cognitive neuroscience, and even more uncertainties about the BOLD brain and cerebral geographies were introduced by the fact that the discovery of false positives was paralleled by the rising prominence of studying what had typically been classified as background noise in brain imaging experiments. Whereas the BOLD response was geared toward contrasting measurements of brain activity "at rest" with brain activity that emerges due to "cognitive load," increased blood oxygenation is now understood less as a "response" and more as an indicator of cognitively significant throughput. Thus measurements of blood oxygenation have not fallen out of fashion; in fact, they are still widely used throughout the field of cognitive neuroscience. However, brain activity that was characterized as noise is now being "recovered" by means of novel experimental paradigms (e.g., see Daunizeau et al. 2011) and alternative statistical techniques (e.g., Smith et al. 2011) that trigger and expose network interactions instead of clearly contrasting local peaks of cerebral energy consumption. The following chapter provides a detailed discussion of how resting state research in brain imaging has helped to reorganize the brain from the inside.

4

False Negatives

A paper published in *The Quarterly Journal of Experimental Psychology* (Sanders et al. 2017) tackles the phenomenon that humans sometimes seem unable to keep unrelated thoughts at bay. In particular, the researchers experimented with techniques to get volunteers out of their heads while reading a piece of text since, as is well known, our thoughts tend to stray from the words on the page as we skim through; what we read calls up memories and brings things to mind that we may have thought about before they got lost in the frenzy of informationally dense environments. Such findings have put brain imaging under considerable pressure: how are we supposed to identify the cerebral correlates of reading, for instance, if volunteers are simultaneously thinking about their life goals, possibly triggered by the experimental task, which has been designed to foreclose such distracted behaviour? What would, in turn, an experimental paradigm look like that allowed us to examine what goes on in our brain when we stray – voluntarily or not – from what we are supposed to concentrate on in our immediate environment?

This chapter takes up the thread of chapter 3, which focused on the critique of false positives in cognitive neuroscience, with an analysis of the discovery of "false negatives" (Lohmann et al. 2017) – or the methodical recovery of brain activity that was inaccurately characterized as pure noise. The underlying so-called resting state activity was disregarded in cognitive neuroscience for decades even though, as I show below, the synchronized activity of large neuronal populations during periods without cognitive input had already appeared in measurements of electrical activity in the brain taken in the early twentieth century.

With new measurement techniques, however, came new epistemologies and experimental paradigms; the electric brain quickly turned into an unknown world that had to be explored and mapped in the first place. Phrenological ideas about discrete brain regions that supposedly govern specific cognitive capacities were revived through techniques to measure local energy consumption in the brain. Brain mapping – or as I call it, cerebral geography – hence focused on contrasting brain activations and disregarded or even discarded brain activity that did not surpass certain statistical thresholds. However, resting state research and "network neuroscience" propagate an entirely different perspective on the brain's workings by conceiving of the brain's "hardware" as no more than enabling resources in network form. These resources appear to be most effectively activated by putting the volunteer in the scanner "to rest."

RESISTING A REST

Before I began to do research on cognitive neuroscience and brain imaging epistemologies, I had come across a number of reports on the experience of "being scanned." In field notes found on the internet or in the literature, former volunteers emphasized the irritating noise of the pulsating helium pump, the claustrophobic atmosphere within the scanner, the fact that their bodies had been stripped to a stretcher and their heads caged into a magnetic coil in order to minimize artefacts in the data, or the strange cognitive tasks they had been administered to over the course of the experiment. My experience was different. Granted, I was not forced to look at anxiety-inducing images; my task was to move a bar that appeared on the screen in front of me up and down, first by squeezing my hands, then by imagining to squeeze my hands.

Yet, in order to discern the neuronal activity that undergirds said task from the so-called baseline activity, I was repeatedly asked to "think about nothing in particular." Instead of calming my brain, said instruction made me aware of my inability to get out of my head. To give you an impression of how enervating an experience this experiment proved to be, I am posting the field notes that I compiled after a long day at the lab below.

I had an appointment with John today, he wanted to take me down to the scanner to witness the calibration of a future

Brain-Computer Interface based on MRI. Romain, a PhD student, was already downstairs and in the process of preparing his experiment. "I am almost set," he explains as we enter the room, "only Chris [the volunteer for the experiment] isn't here yet." While we are waiting, he discusses details of the experiment with [his colleague] Jonas. Thirty minutes go by, but Chris remains a no-show. "Man, I think he forgot it again, he's not very reliable, you know, tends to forget these things," Romain explains. "Which is bad for us, scanner time is precious," he adds, and turns to me: "Would you want to do it? It's simple, it's just for the calibration of our classifier. Do you have a pacemaker? Any piercings you can't take out? Does the ink of your tattoos contain iron?"

I am not good at saying no and it certainly gets worse if people catch me off guard. Some minutes later, I am standing there, without a shirt, discussing my fitness, as if I were back at physical examination for military service. "Should you have the feeling that your skin is heating up, immediately press the button. Don't wait till it smells like barbecue," Romain adds with a cheeky smile that makes me even more nervous. What if it really starts to burn? It's that feeling that you have when you're at the doctor's for a haemogram, anticipating a stabbing pain although you would typically barely notice a sting.

Standing in front of the heavy door to the scanner room I am asked to empty my pockets: "Please dig deep. We're maybe a bit paranoid, but recently someone forgot a coin in his pocket, it was forced out by the magnet and destroyed the screen inside the scanner." He wouldn't have had to tell me: the door to the scanner room is plastered with warnings – no pacemakers, no metallic implants, no loose ferromagnetic objects etc. – and I had been staring at it for accumulated hours already. Bold red letters reading: "A n'ouvrir qu'en cas de danger de mort" don't really calm me down.

From inside the scanner room, the helium pump that cools the scanner's magnet sends loud pulses. I am asked to lie down on the stretcher. Thomas, the PhD student, pulls hearing protection over my ears and wraps my head in a sort of cage that supposedly increases the strength of the signal. By means of a small mirror mounted above my head I can see into the gaping hole behind me. Slowly the stretcher moves backwards into the scanner until a cross of light hits my forehead. Adjusted and ready to be scanned!

"We'll show you some random images now," Romain explains over the intercom, "just watch them go by and relax. We're trying to get at the baseline of your brain activity to train the classifier. Just try and think about nothing in particular." The Taj Mahal appears on the screen, followed by a group of monkeys delousing each other, a sundown on some remote, palm-fringed beach and so on it goes. Pictures appear and disappear. The rhythmic pulses of the helium pump, the images, the position – I start to get dizzy and need to force myself to stay awake. My mind begins to wander and as I notice, I can't help and contemplate about the effects of the wandering mind on the experiment, and then about the effects of contemplating on my brain activity, and about how strange a concept contemplation is anyway.

I don't know a lot about neuroscience, but "thinking about nothing in particular" is probably not what I am doing right now. If I am thinking about this right now, am I ruining the experiment? I mean, they're trusting me to perform, but whatever I am trying to do, it seems to make things worse. Thinking about what "rest" really means, how in the world could I ever put this brain to rest?

That my brain successfully resisted my attempts to put it to rest illustrates how industrious and cognitively rich the brain's resting state indeed is. Nevertheless, it had in traditional brain imaging paradigms been treated as a mere baseline activity that could be classified and discarded as pure noise.

In an article entitled "Does the Brain Have a Baseline? Why We Should Be Resisting a Rest" (2007), published by the influential journal *NeuroImage*, Alexa Morcom and Paul Fletcher of the Brain Mapping Unit at the University of Cambridge react to the growing importance of resting state studies within the field of imaging neuroscience. The idea that this resting state might represent a "default mode" of the human brain, they argue, is not justified simply by "observations that a consistent network of brain regions shows high levels of activity when no explicit task is performed and participants are asked simply to rest" (1073). Although a resting subject's brain might well be active, the authors see no reason why such activity ought to be conceived as a highly subjective and subject-constituting brain state.

Morcom and Fletcher's (2007) resistance to a reification of the brain at rest as a distinct "state" must be understood in the context of cognitive neuroscience methodology and the BOLD paradigm, where brain activity unrelated to an experimental stimulus or task has been regarded – or better, disregarded – as "noise." Within this paradigm, the so-called baseline activity, or "control state," is recorded independently of and subsequently subtracted from the "task condition" in order to isolate regions where activity increases in response to a stimulus or an experimental task administered to the volunteer in the scanner. To be classified as meaningful cognitive activity, the accumulated activity of neuronal populations needs to surpass a certain threshold so that an entire brain region stands out from the neuronal crowd. Cognitive neuroscientists thus speak of "contrast" when they refer to strategies of highlighting areas of increased activity (e.g., see Logothetis and Pfeuffer 2004).

Over the course of the 1990s, however, the interest in "resting state activity" grew as observations of an *increase* in regional brain activity during the control condition of brain imaging experiments became ever more frequent. Bharat Biswal and colleagues (1995) were the first to perform an analysis of brain activity that occurred *after* the actual experimental task, finding that volunteers did indeed show cognitive activity even though their brains were purportedly "at rest." Two years later, a group of researchers at Washington University noted an actual *decrease* of activity in some brain regions while volunteers performed an experimental task. Deciding to dig deeper, Gordon Shulman and colleagues (1997) subtracted the task condition from the resting condition (instead of the usual subtraction of the resting state from the task-related activity). Thanks to this simple "flipping of contrast" (Callard and Margulies 2011, 233),[1] they found that some regions in fact showed higher activity at times of "rest" than during goal-directed task conditions (see also Mazoyer et al. 2001).

Around the turn of the twentieth century, the ranks of those who critiqued the marginalization of brain activity that fell below the statistical threshold of cognitive neuroscience were swelling. As the "resting brain" suddenly appeared to be highly active with the advent of resting state research, attention was diverted to brain activity that had previously been experimentally and statistically muted. Relying on Rodolfo Llinás's (2001) genealogy of how intrinsic brain activity has been strengthened at the expense of a "reflexology" (98), Marcus Raichle and colleagues (2001), at the forefront of the so-called resting state studies, explain how the novel paradigm of cognitive neuroscience might best be conceived. Whereas brain reflexology deems the inner workings of the brain significant only if they can be reliably linked to experiment, resting state research explicitly declares the opposite to be its research object: activity that is "intrinsically" motivated because it occurs in the absence of external cognitive stimuli.

Debra Gusnard and Marcus Raichle (2001), for instance, suggest that intrinsic activity might be substantial and functionally important in the construction and maintenance of human subjectivity. For many neuroscientists, however, such ideas appear to be out of sync with the cognitive neuroscience reality. As Morcom and Fletcher (2007) argue, it seems highly unreasonable to expect that having research subjects lie on their back and do nothing will help with the development of a distinct personality profile. However, such criticism is primarily based on the notion that resting state imaging is representative of no more than laziness on the side of the experimenter, who thus elicits certain brain responses by merely bypassing the highly sensitive process of "configuring" experimental subjects.

The necessity of carefully designed experiments has always been a strong argument against outsourcing fMR imaging to decentralized observatories and data collection facilities.[2] Another of my interviewees, George, the director of a leading British brain imaging lab, addressed this issue in our conversation about the feasibility of long-distance collaboration in the neurosciences, taking the European Organization for Nuclear Research (CERN) as an example:

> [T]he technology is so nose-bleedingly expensive that that's driven the concentration of physics researchers into interacting teams of tens and hundreds who are dominant in the field. You can't do particle physics without going to CERN! Where functional imaging hasn't quite crossed that bridge is that the technology isn't so expensive that it had to have one place in the world or two places in the world that do it. And so you haven't had that transition, I think, to a model where, like with physicists, where you would go to CERN and do your observation and then take your data away, and that would be five years' worth of data. Some people had that model, though. I was a postdoc at CalTech in the late '90s, and they tried to set up their imaging model on the basis of their observatories, which were very much a technical piece of equipment, you know, some curator looked after. And clever scientists came

in, did their observing, and then retreated with their data and did data analysis. And that model hasn't worked for them [since] they couldn't set up their centre like that, and it hasn't worked for imaging in general.

Preparing a classic brain imaging experiment within cognitive neuroscience involves a volunteer who is calm and purportedly cognitively inactive, disengaged from the experimenter, and waiting to be instructed as she lies in the scanner. Meanwhile, the experimenters are seated in the control room and calibrating their instruments while (secretly) recording the activity of the volunteer's putatively unaffected brain. The experiment officially begins as soon as a baseline of brain activity has been established, and it involves cognitive or mental tasks designed to selectively activate the brain in contrast to its so-called resting state, which "may persist through both experimental and rest epochs if the experiment is not sufficiently challenging" (Greicius and Menon 2004, 1484).

One of my interviewees, Ethan, a postdoctoral researcher employed at a British brain imaging lab, explained to me that differentiating between brain states is first and foremost a challenge of triggering certain activations and avoiding others. Researchers can either trust the experimental subject to perform as expected or actively work around her experimental agency. Experimenters often try to enforce the volunteer's compliance by forging brief yet intimate personal relationships, attempt to minimize individual agency via experimental design, or even trick their subjects into behaving as desired (Cohn 2008). Such strategies seem necessary, as one brain imager told anthropologist Simon Cohn (2004), since "you can just never be sure that they're not thinking about the weekend, or something" (64). Task-based brain imaging is thus essential to designing an experiment in such a way as to exclude everything that might water down the targeted brain state. At the same time, contrast can be increased by designing the resting state itself; for instance, in the experiment mentioned at the beginning of this chapter, rest – or the control condition – involved visual stimuli that were selected because they were entirely unrelated to the goals of the experiment.

No matter how exactly the experiments of cognitive neuroscience were designed throughout the twentieth century, they focused almost exclusively on activations instead of activity. The brain's resting state, for instance, did not appear to constitute a clearly distinguished cognitive mode. Defined simply as the obverse of an (experimentally) activated brain and designed as an experimental control condition that promoted the absence of cognitive activity, the resting state seemed entirely relational, an artefact of brain imaging's traditional experimental design. The proponents of resting state studies, however, made a virtue of the seemingly inconsistent character of the resting state and argued that "any control state, no matter how carefully it is selected, is just another task state with its own unique areas of activation" (Raichle et al. 2001, 676). After all, the brain's level of activity "during 'at rest' states will inevitably depend on the nature of the experimental task condition to which the resting state is compared" (Callard, Smallwood, and Margulies 2012, 3).

In experiments reported by Raichle and colleagues (2001), the resting brain appeared entirely unrestful and highly active. They consequently argued that changes in activity measured in activation-based brain imaging experiments represent but the "tip of an iceberg" (Morcom and Fletcher 2007, 1073) since the by far greater part of consciousness-generating activity is drowned in cognitive neuroscience's very own ocean of noise. The possible cognitive significance of the noisy activity of neuronal populations was already hypothesized by researchers prior to the 1930s. As the next section shows, however, positron-emission tomography (PET) and fMRI seemed to lend themselves well to the task of localizing cognitive activity and thus to the goal of functionally segregating the "space inside the brain" (Beaulieu 2002, 17).

NOISY CROWDS AND PHYSIOLOGICAL LANDSCAPES

Something akin to resting state activity had surfaced around 1930, after Hans Berger (1929), director of the Psychiatrische Universitätsklinik in Jena, Germany, published the results of a long series of experiments with wave-like signals in the brains of animals and humans. His publication summarized the results of nearly thirty years of noninvasive explorations of human brain function in vivo with early versions of the electroencephalograph (EEG). Berger had shunned the then dominant method of recording the activity of nerve cells during, for instance, open brain surgery and opted for registering the brain's accumulated activity by attaching electrodes to the bare skin of the skull. Indeed, Berger observed that whenever he placed the electrodes of a sensitive ammeter on the head of a subject, two distinct types of waves – later called alpha and beta waves – occurred largely independent of pulse, heartbeat, muscle activity, and respiration. Searching for coherent curves or patterns in the graphic notations of the EEG, Berger was able to stabilize wave-like signals that he interpreted as surface representations of intrinsic brain activity. "Mental work," Berger (1969) explained, "adds only a small increment to the cortical work which is going on continuously and not only in the waking state" (563).

In the context of neurophysiology's analysis of communication processes among single nerve cells, however, Berger's experiments appeared to be inherently reductionist and were interpreted as a lack of appreciation for the enormous complexity of the brain. His discovery was barely referred to until 1934, when British Nobel Prize laureate Edgar Douglas Adrian and his assistant Rachel Matthews repeated Berger's experiment in search of an explanation for disturbances that had occurred in their lab's analyses of nerve fibre action potentials (Borck 2005). Interestingly, what they referred to as disturbance was an emerging regularity in otherwise irregularly occurring "brief explosive waves" (Adrian 1965, 297).

In an article on experiments involving the stimulation of the eye of an eel with light pulses, Adrian and Matthews (1928) mentioned the regular waves for the first time: "When the entire retina of the Conger eel is exposed to uniform illumination the action current discharge in the optic nerve may lose its usual irregular character and may consist of a series of regular waves with a frequency which usually varies between 15 and 5 a sec" (296). Similar "disturbances" were found in the brain of the goldfish (Adrian and Buytendijk 1931) and in the ganglia of the water beetle (Adrian 1931, 1932).

In search of an explanation for the disturbing regularity, Adrian and his assistant Bryan Matthews used Berger's method and successfully reproduced the alpha rhythm, which they subsequently showcased during a conference of the Physiological Society in Cambridge, England. The British cyberneticist William Grey Walter (1934) reported their experimental production of cortical potentials in an article for the *Spectator*:

Adrian and Matthews recently gave an elegant demonstration
of these cortical potentials ... When the subject's eyes were open
the line was irregular, but when his eyes were shut it showed

a regular series of large waves occurring at about ten a second
... Then came the surprise. When the subject shut his eyes and
was given a simple problem in mental arithmetic, as long as
he was working it out the waves were absent and the line was
irregular, as when his eyes were open. When he had solved
the problem, the waves reappeared ... So, with this technique,
thought would seem to be a negative sort of thing: a breaking of
the synchronized activity of enormous numbers of cells into an
individualized working. (479)

Historian of science Cornelius Borck (2008) argues that the
discovery of unconstrained activity effected contemporary understandings of the brain in general, with the result that "the central
nervous system was no longer [regarded as] a central telegraphy
office processing incoming messages and outgoing commands, but
a strikingly active source of intrinsic activity calling for new inter-
pretive metaphors" (371). Indeed, the EEG enabled and gave rise to
distinct interpretations of alpha waves, ranging from mathematical
analyses aimed at finding a universal brain code (Wiener 1957) to
pictorial interpretations of lines and curves on paper in an attempt
to draw distinct images of disorder (Gibbs and Gibbs 1941).

Berger's search for regularity in the otherwise chaotic noise of the
electroencephalographic brain testifies to a technologically defined
interest in fluctuating potentials and rhythms, which turns the
graphic inscriptions of brain waves into an early means of delineating intrinsic activity from task-related activity. However, by visually
aligning his own alpha waves to regular signals emitting from the
water beetle's ganglia in the absence of light, Adrian strengthened
a line of thought that equated cognitive activity with a response to
the demands of the outside world vis-à-vis interpretations that, like
Walter's, suspected intrinsic, cognitive activity.

Adrian's reinterpretation of EEG measurements marked the beginning of a paradigm that focused on brain *activations* due to external
stimuli instead of looking at its relentless activity. Although novel
technologies such as positron-emission tomography provided a
potential register for the observation and quantification of brain
activity, its measurements were instead aligned with the goals of an
impending brain mapping community. At the National Institutes
of Health in the United States, physiologist William Landau and
colleagues (1955) succeeded in linking in-vivo measurements of

changes in regional blood flow in the brains of laboratory animals to physiological activity, but members of the group remained skeptical about the significance of their own recordings. Landau, for instance, offered a very cautious characterization of their autoradiographic measurements and pre-emptively disqualified the results as purely instrumental: "Of course we recognize that this is a very secondhand way of determining physiological activity; it is rather like trying to measure what a factory does by measuring the intake of water and the output of sewage. This is only a problem of plumbing and only secondary inferences can be made about function" (128).

Landau's interpretation of the groups' experiments as measurements of intakes and effluents testifies to an interpretation where increases and decreases in blood flow were regarded as infrastructural activity that must not be interpreted as an indicator of (intrinsic) cognitive activity. Even after positron-emission tomography had been approved as a means to "image" brain activity, it bore traces of a second-hand approach to the workings of the brain. Its intriguing novelty was bound to the visualizations that it helped to produce, for even though PET's "metabolic landscapes" (Beaulieu 2002, 17) proved rather hard to interpret, their rough resemblance to brain-shaped heat maps appealed to the eyes of early neuroscientists.[3] Attempts to map these physiological landscapes gave rise to the desire to standardize the terrain they occupied; after all, the coarse shapes of PET's landscapes neither provided reliable information on brain structure nor accounted for the variability of individual brains (Kennedy 1991).

In the 1980s researchers at the University of California in Los Angeles and at Washington University in St Louis developed a method for allotting the metabolic landscapes of PET to a grid of coordinates, which was to be superimposed on anatomical images generated by means of magnetic resonance or computer tomography (Fox, Perlmutter, and Raichle 1984; Mazziotta et al. 1982). The initial success of this method, again, pushed the development of a functional version of MRI, which promised to allow researchers to align measurements of activity and images of brain structure in a single apparatus.

The breakthrough for functional MRI came in the early 1990s, when a group at AT&T Bell Laboratories succeeded in producing images of blood flow by means of magnetic fields of varying strength. Seiji Ogawa and colleagues (1990) were experimenting with different levels of oxygen concentration in the air that rats were breathing when they observed that a high oxygen level made blood vessels in the brain disappear from their scans. Ogawa and colleagues determined that the vessels' disappearance resulted from a decrease in the magnetic properties of blood when haemoglobin is oxygenated and therefore called it blood oxygen level–dependent (BOLD) contrast. The authors maintained that "BOLD contrast adds an additional feature to magnetic resonance imaging and complements other techniques that are attempting to provide positron emission tomography–like measurements related to *regional* neural activity" (9868, original emphasis).

The brain's restless activity was consequently defined as a "baseline" that needs to be subtracted from the task condition in order to etch meaningful activity into relief, and brain imaging came to impart a novel empiricism to an older epistemology represented by Korbinian Brodmann's (1909) localization of brain areas on the basis of his investigations into the cytoarchitecture and myeloarchitecture of brain cells. Gareth recounted the introduction of BOLD imaging as the moment when, "for the first time, you could confirm the suspicion that the brain was actually very delicately and very systematically organized in its computational abilities and show this functional segregation. It was the excitement of discovering a new continent, and the first thing you do when you discover a new continent is what? You map it. You're a cartographer."

Originating in the realm of anatomy and providing relatively fine-grained spatial resolution, MRI lent itself well to observing local increases in energy consumption and was consequently wed to the experimental paradigm that qualified PET and fMRI as techniques for the observation of brain activations in response to external stimuli.

CEREBRAL GEOGRAPHY I: TERRITORIES OF THE MIND

Although imaging brains has always been a local affair, the comparability of individual brain scans had to be established through centralized efforts. The decisive step in the production of comprehensive maps and atlases was the development of an intersubjective coordinate system that allowed researchers to align individual brains, define shared regions based on the brain's anatomy, and develop statistical techniques to analyze activity in "regions of interest" to the experimenter. In order to map peaks in energy consumption as clearly segregated "regions," a multitude of brains first had to be forced into a common reality. Neuroscientists developed a wide range of templates from the brains of single subjects and pre-labelled brain regions in accordance with the infamous Talairach atlas, (Talairach and Szikla 1967; see also Chau and McIntosh 2005), which is based on the postmortem dissection of a sixty-year-old French female's brain. The International Consortium of Human Brain Mapping[4] later took on the challenge of developing a multisubject template that was supposed to represent a "normal" and "average" brain, which obscures the variance of individual brains to a certain degree and turns the remaining differences into features of a population brain.[5]

. In a programmatic article, the consortium's principle investigators described the challenges of establishing the population brain as follows:

> Earth atlases can assume a relatively constant physical representation over thousands of years. On that single, stable construct, an infinite number of abstract representations of features can be overlaid. For earth maps, such features might include rainfall, temperature, population density or crime rates. Unlike geographical atlases, anatomical atlases cannot assume a single, constant physical reality ... Anatomical atlases must first deal with the fact that there is a potentially infinite number of physical realities that must be modelled to obtain an accurate, probabilistic representation of the entire population. Upon this anatomical representation one can then overlay features in a fashion analogous to that described for earth atlases. (Mazziotta et al. 2001, 1294)

In order to turn an infinite number of physical realities into a more or less stable, biophysical reality of *the* brain, cerebral geographers have put a comprehensive system of statistical techniques in place. The data of individual brain scans first need to be pre-processed and "registered" on a template. Figure 4.1 shows an illustration of the entire experimental process, where spatial normalization is part of quality control and pre-processing. A minimal set of pre-processing steps includes corrections for head movement that might occur during the scanning process and the "warping" of an individual brain into a common brain space. A spatial grid helps to adjust the size of the

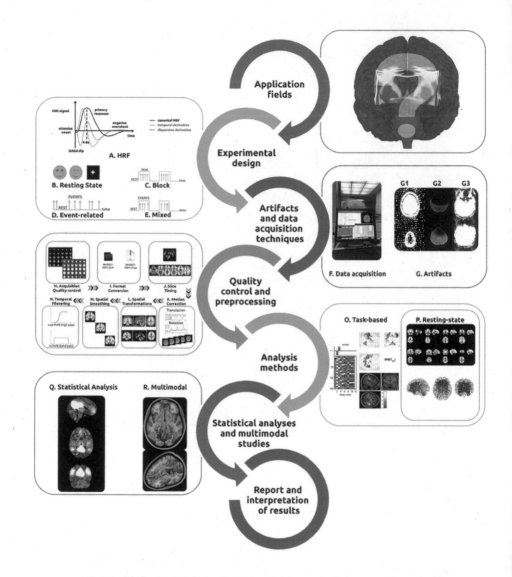

4.1 "A Hitchhiker's Guide to Functional Magnetic Resonance Imaging" (Soares et al. 2016).

individual brain so that it fits the template, which is then used to define functional regions in the individual brain. In a further step, a procedure called "spatial smoothing" corrects for anatomical differences between the individual brain and the population brain. After normalization, the data therefore are effectively derived from a mixture of individual brain and templates that contain a few healthy, well-educated, and predominantly white brains.[6] The population brain of the atlas is, so to speak, averaged into existence.[7]

In order to compare the brain activity of individuals, the procedure is repeated whenever a new brain is scanned and "registered" on a template. Yet activation data must be averaged and cleaned of artefacts as well. In this process, various parameters may be tweaked at any time in the interest of producing better images, whereupon "better," whose definition is at the discretion of the researcher, determines "researcher degrees of freedom" and might simply mean that the images are more compelling (Gelman 2016; see also Dumit 2011). This is to say that the preparation of significant activations *after the fact* is the most significant step in the process and relies on increasing the contrast of a targeted brain response. Changes in the MRI signal are characteristically minimal, which is why the blinding "forest fires" that we are used to seeing sparsely distributed over the predominantly grey matter of a brain purported to be only selectively activated appear most distinctly by minimizing the effect of those populations of neurons that are not "on fire."[8] This step may involve (visually) emphasizing brain regions or networks that show heightened activity when the volunteer completes a task in the scanner and are therefore identified as significant *after* the experiment. Increasing the contrast might also relate to defining regions of interest, such as the hippocampus, *before* the experiment as a means to show in what way these are involved in a specific task.

The BOLD response generally relies on an all-or-nothing principle of activation: the neurons contained in a volumetric pixel element (voxel) must be active on average if they are to surpass the threshold and light up within a terrain that looks like it is devoid of any activity and is accordingly dark. A predefined or experimentally adjusted threshold determines whether regions appear activated and "stand out from the crowd" at the expense of activity that is then characterized as pure noise. Adjustments can generally be made at any stage, yet the process is at least partly automated through software tools, which are programmed and maintained by methodologists. It might not be surprising, therefore, that "significant" brain activity is often disregarded by accident – or methodically discarded – within a paradigm that regards only the most contrasting activations as significant and has therefore helped to coin our image of an only sparsely activated human brain.

Whereas the voodoo controversy exposed how and to what extent these sparsely distributed activations had been inflated, inquiries into the intricacies of statistics undergirding the BOLD paradigm provide insights into the extent of "false negatives." Gonzalez-Castillo and colleagues (2012) have found that increasing the duration or the frequency of scans within scanning sessions has more or less the same effect as greater spatial resolution, allowing them to map activity in 70–90 per cent of the brain's functional areas while the volunteer in the scanner performs cognitive tasks. Johannes Stelzer and colleagues (2014), a team of researchers at the Max-Planck Institute in Leipzig and Tübingen, have concluded that virtually every fMRI may have overlooked the involvement of many significant brain areas due to insufficient spatial resolution. They relied in part on data from fMRI scanners with a magnetic field strength of 7 Tesla instead of the still often used 3 Tesla. The higher strength provides greater spatial resolution and allows for a more detailed mapping of functional brain structures. As they state, "[I]f it is true that even for a simple task the entire brain becomes involved, then the dichotomy of labelling brain areas as active or inactive is no longer meaningful and scientifically relevant. To put it pointedly, these results may herald the end of *qualitative* activation-based neuroimaging" (11, original emphasis).

CEREBRAL GEOGRAPHY II: NETWORK NEUROSCIENCE AND THE DEFAULT MODE

Whether approached through increasing spatial resolution or through a more fine-grained temporal resolution, it is increasingly clear that the sparsely activated and neatly segregated brain exists only on the statistical level – being mediated by minimal peaks in the brain's energy consumption that are statistically isolated to be depicted in brain maps or atlases as regions "on fire." Shifts in the use of statistical techniques call the entire endeavour of a static, activation-focused cerebral geography into question. The insight that virtually "all domains of cognitive function require the integration of distributed neural activity" (van den Heuvel and Sporns 2013, 683) has pushed research agendas that aim to analyze how the brain might be organized into complex networks and "hubs," which channel the flow of information throughout the entire brain space.[9] Contrasting peaks in energy consumption are thus substituted with the macroscopic patterns of neural activity that, emerging from cognitive neuroscience's data shadows, have brought a shadow system to the fore, which has literally put flesh on the notion of an unstable resting state.

Since the mid-1990s, the default network has pushed its way into the limelight of contemporary cognitive neuroscience; although the term did not appear among the ten most used within the neurosciences until the year 2010, it made the list four times in a row and came in fourth in 2015, outranking – savour this! – the concept of functional connectivity as well as the very technique that brought it into being: neuroimaging (Yeung, Goto, and Leung 2017).

In contempt of Morcom and Fletcher's (2007) critique of resting state studies, neuroscientists did not refrain from analyzing brain imaging's baseline condition but instead transformed it, first, into a specific brain state that occurs when cognitive input is absent and, subsequently, into a mode of brain function that has its very own neural correlates. This trajectory from resting state to default network involved various analogies and metaphors, the most significant of which was the "dark energy" metaphor, which initially linked the brain's resting state activity to the roughly 70 per cent of unaccounted for energy that permeates the universe (Raichle 2006).[10] Just as in the realm of physics, the dark energy metaphor was chosen as an auxiliary hypothesis to explain the occurrence of a phenomenon that cannot be accounted for within the existing theories of brain function. Dongyang Zhang and Marcus Raichle (2010), for instance, argue that the "visible" elements of brain activity account for less than 5 per cent of the brain's energy budget, overshadowing the by far greater part of significant neuronal chatter, or the "dark energy" of the brain.

To put it differently, the brain's resting state was thus turned into "dark energy" since it was derived from observations that threatened to falsify the dominant paradigm within the field of cognitive neuroscience (Raichle 2010).[11] What we now know as the default mode of brain function crept out of the "data shadows" of cognitive neuroscience – emerging from data that were missing, incomplete, ignored, unwanted, or characterized as unreliable (Leonelli, Rappert, and Davies 2017). What had been disregarded within the traditional BOLD paradigm as "false negatives," if you will, supported the appearance of a coordinated mode of brain function that had previously been unheard of.

Indeed, evidence that the resting state might constitute only an unorganized and noisy control condition had accumulated since at least the time of Nancy Andreasen and colleagues' (1995) observation that the brain's resting state "is in fact quite vigorous and consists of a mixture of freely wandering past recollection, plans for the future, and other personal thoughts and experiences" (1578). Leonard Giambra (1995) was the first to use the term "default mode," having concluded that "task-unrelated imagery and thought ... may represent the normal default mode of operation of the self-aware" (1). However, the term did not come to real prominence until Debra Gusnard and colleagues (2001) described it as an "organized mode of brain function" (4259) that is suspended during goal- and task-directed behaviour.

In an interview with *The Meditation Blog*, Raichle explains the discovery of the so-called default mode network through the analysis of data that had been lost to the resting state:

> We discovered the default mode network when we asked participants in a study to perform tasks that were so demanding that they had to be absorbed in what they were doing. It is well known that when involved in such tasks, we use the attention network of the cortex. We noticed that activity in certain areas of the cortex was reduced during the involvement in demanding tasks. It was really surprising that, after the demanding tasks were completed, activity in these areas of the cortex increased again. The brain seemed to revert back to a default activity level, which is there in the absence of a specific, ongoing, external task. (Quoted in Davanger 2015)

Although the instruments had of course changed, note the striking continuity between this account and the demonstration of intrinsic activity in Edgar Douglas Adrian's experiments in Cambridge as described by William Grey Walter (1934). This time, however, the conspicuous, rhythmic fluctuations of cognitive activity did not appear as amplitudes in recordings of neuronal chatter but as spatially distributed oscillations in brain activity and could therefore be reconciled with the still dominant geographical paradigm of brain imaging through the notion of an underlying network of brain regions that are typically involved in the default mode (Greicius et al. 2003). Leading neuroscientist Giorgio Tononi circumscribes the emergence of the default mode network as "the discovery of a major system within the brain, an organ within an organ, that hid for decades right before our eyes" (quoted in Callard and Margulies 2010, 341). As Michel Serres (1982) puts it, "[O]rder sometimes comes only from an explosion of noise" (20).

Once the default mode network was etched into relief, experiments with resting brains in scanners gradually revealed its cognitive significance and its connections to other large-scale brain networks. Whereas most researchers initially conceived the default network as anti-correlated with other large-scale yet task-specific networks (see also Fox et al. 2015; Raichle 2015a), others swiftly cast shadows of doubt over the purportedly strict coupling of cognitive rest and the default network. Randy Buckner, Jessica Andrews-Hanna, and Daniel L. Schacter (2008), for instance, argue that the "default network's prominent use during passive epochs may contribute adaptive function by allowing event scenarios to be constructed, replayed, and explored to enrich the remnants of past events in order to derive expectations about the future. This functional role may explain why the default network increases its activity during passive moments when the demands for processing external information are minimal. Rather than let the moments pass with idle brain activity, we capitalize on them to consolidate past experience in ways that are adaptive for our future needs" (31).

From this perspective, the alleged antagonism between the default network and executive functions appears to be a result of a competition for limited resources between external and internal demands rather than a result of their functional incompatibility (Smallwood and Andrews-Hanna 2013). Indeed, the default network is co-activated with the "executive control network" and the "salience network," as shown by studies, for instance, of creative incubation (Beaty et al. 2015) and autobiographical planning (Gerlach et al. 2014; Spreng et al. 2015). It has therefore come to connect – and this outcome is highly remarkable as regards twentieth-century cognitive science and psychology – goal-directed action with the "flights of fancy" that we have traditionally understood as distracted, "self-generated," or "stimulus-independent" thought (Mason et al. 2007).

Within a decade, rest was transformed from a state of cognitive inactivity into a mode of brain function that is considered to be the source of creativity, sociality, and subjectivity. Resting state imaging is therefore on par with the kind of activation-based imaging characteristic of twentieth-century cognitive neuroscience and extensively used to produce data of dynamic interactions between large-scale brain networks with a greater temporal resolution. Complicated mental tasks are used as an alternative to "thinking about nothing in particular," and said experimental reconceptualization bears witness to the fact that spontaneously or intrinsically generated activity is considered to expose the brain's information processing capacities most effectively. Although the methodology of resting state imaging still bears its original name, the former resting state now figures as a form of industriousness, clearly dissociated from what we would typically understand as "slacking."

The title of Raichle's article "The Restless Brain: How Intrinsic Activity Organizes Brain Function" (2015) is consistent with Antonella Marchetti and colleagues' (2015) view that the default mode is "the mind, which does not rest even when explicitly asked to do so" (1). Both make a statement about the brain that seems to capture how the dark energy of the brain has been relentlessly destabilizing cognitive neuroscience's functionally segregated brain from the inside. The plasticity of the default network – as regards both the exact neuronal populations it involves and the cognitive processes it is recruited for – has contributed to shifting the focus of many neuroscientists away from an interest in the functions of specific brain regions and toward the investigation of highly complex cognitive phenomena in terms of their infrastructure, which Geoffrey Bowker and colleagues (2010, 98) refer to as *pervasive enabling resources in network form* (original emphasis).

(IN)FORMATION

In 2010 social scientists Geoffrey Bowker, Karen Baker, Florence Millerand, and David Ribes called for studying ways of knowing in networked environments through an analysis of information infrastructure where "conventional understandings of infrastructure as 'tubes and wires' [are extended] to the technologies and organizations which enable knowledge work" (Bowker et al. 2010, 98). The authors describe a novel mode of infrastructure studies where physical components and seemingly homogeneous entities are analyzed in terms of their function within an overarching system that is much more fuzzy and frayed than its elements might suggest. This approach allows us to gain a better understanding of the whole, but it might also help in reconceiving the constitutive elements that form part of the system.

This was the work of the default mode network. Its emergence from shadow data provided cognitive neuroscience with an entirely different perspective on the brain, for it came to constitute the infrastructure of the social and creative, human individual. "It may be off when you're on, but the brain network behind daydreams and a sense of self is no slacker," Tina Hesman Saey writes in an article titled "You Are Who You Are by Default" (2009), where she emphasizes how distracted states of mind were dissociated from the realm of psychopathology as soon as they could be conceived as a neuronal network.

Network neuroscience projected said understanding of brain organization onto the entire system and turned the default network into a sort of interscalar vehicle: an object that became a mode of analysis and allowed researchers to move freely through different spaces and perspectives (Hecht 2018). In the case of the brain, the territory of the mind has been turned into an infrastructural space, and mapping the world of the brain has been complemented or even substituted by the quantification and analysis of network traffic throughout the brain. It is fair to say, therefore, that the default network has contributed to reorganizing cognitive neuroscience's brain from the inside, reconceiving it as a highly plastic and ultimately indeterminate assemblage of information highways, switches, and hubs.

As much as traditional cerebral geography relied on peaks in energy consumption that were only statistically present, the novel complexity of brain organization is statistically enacted through multivariate pattern analysis that focuses on the temporal dynamics of large-scale networks or, put differently, fluctuations in neural network activity (e.g., see Van Calster et al. 2017) Mapping continues to play an important part in brain imaging – as regards "connectomes," for instance, which trace the physical pathways of communication between distributed neuronal populations. The gaze of computational approaches to cognitive neuroscience, however, has gradually veered away from the brain's hardwired structure. Just as undersea cables sink to the bottom of the oceans that they are supposed to bridge, the brain's hardwired structure increasingly fades into an invisible background when the analysis of network traffic sits at the centre of attention. Whether it is transmitted through nerve fibres or optical cables, information knows no borders – and increasingly not even the substance through which it flows.

To a certain degree, contemporary brain imaging latches onto the metabolic landscapes of early PET, which now appear, instead, as iridescent cerebral "energy landscapes" (Gu et al. 2018). At the same time, however, brain imaging's brain increasingly invokes geographies of information and resembles the capacious figure that has come to denominate and represent planetary-scale information infrastructure: the infamous Cloud.

5

Cloud Geographies

Merely obtaining new knowledge about digital culture's materiality
may not address the root problem. That problem, as I see it, has to do
with our mental map of cloud computing, the heuristic that we use to
imagine how information is organized, whether in physical space or in
digital space. Recall that cloud-computing software maps a common
infrastructure into individual users.

Tung-Hui Hu, *A Prehistory of the Cloud*, 2015

During one of our conversations, Gareth had to endure my more or
less woolly thoughts about current trajectories in cognitive neuroscience, brain imaging, and indeed the brain itself. Instead of formulat-
ing precise questions, I was tossing metaphors and analogies at him
in an attempt to get to the point, when he suddenly leaned forward
and interrupted me:

It's more the configuring and understanding and maintenance,
well, not maintenance, but certainly the understanding of the
communication [between neuronal populations], the globaliza-
tion of it. So it's like in information technology. It's not so much
about what is connected; it's how it's connected and how that
connection is managed. So it would be the difference of getting
excited about the invention of the first computer and actually
having a thing that was working to the perspective we have now
where the actual implementation is completely irrelevant. It can
be in the Cloud, you know. It doesn't matter.

What Gareth describes here is a trajectory that I have tried to
approach through a discussion of changes in brain imaging methodology throughout the previous two chapters. His account explains,
from a different perspective, why any geographical reference or

analogy will have to be discouraged if we are to develop a more useful and, as regards modelling brain function, more "functional" understanding of how the brain does what it does. Whereas I have described how a different approach to the use of statistics contributed to the idea of highly complex and somewhat unpredictable brain, Gareth alluded to a related but ultimately distinct idea: that the biophysical reality of the brain and all the processes that contribute to its functioning might not need to be comprehensively understood in order for us to model and re-engineer higher cognitive functions. "The exciting thing," he elaborated, "is the graph theory treatment of the way it is organized and the way that you actually exchange information between the nodes of computation."

Newer, computational approaches to cognitive neuroscience tend to disregard the biological implementation of cognition in order to increase researchers' degrees of freedom in modelling the specific *employment* of cognitive infrastructures distributed throughout the space inside the brain. Graph theory is one of the central tools used to help us unlearn the idea of the brain depicted in atlases and maps in favour of its "infrastructuralization." This process preliminarily crystallized in a controversy that ensued as the first review phase approached for the Human Brain Project (HBP), which at the time was one of the European Commission's two Future and Emerging Technologies Flagship projects, funded with up to 1 billion euro. Neuroscientists passionately discussed the contentious issue of whether the HBP's approach to understanding brain function would be worth the while and, more critically, the money. Why simulate a whole human brain in silico if the neurobiological parameters of brain function are still largely unknown? Neuroscience-inspired AI is set to rely on imaging paradigms or geographies that are very different from those based in fantasies of comprehensive atlases and categorization.

GRAPH THEORY

"What the heck is graph theory?" my colleague asked, turning to me with a skeptical look after he had skimmed through an excerpt of this chapter. Initially, I too had trouble understanding its significance. Graph theory was repeatedly mentioned in many conversations I had with neuroscientists, although it seemed to defy the use of the term "theory." In essence, graph theory appeared to me as no more than a complicating term for thinking about and representing relations in terms of nodes and edges, which implies that a range of entities are considered to be connected and interacting in a specific fashion.

The significance of graph theory can be understood only if it is conceived as a tool. Characteristically unspecific, it is rarely considered a research object of its own. Its various incarnations are tossed back and forth between diverse fields, where they are modified for application to specific research objects. As a tool, graph theory acquired its prominence in the domain of data analysis, for it is the tool that is used in the analysis of many forms of currently social interaction. Graph theory, as it were, *is* the science of networks.

Anna Munster (2013) observes that a "network image can be rendered for just about every aspect of day-to-day life for which data exist, including financial information, organizational data, mapping systems of every variety, social networks, and technical ecologies of all kinds" (2). Indeed, graph theory has turned into a paradigmatic form of representing real-world networks, regardless of whether they are composed of molecules, neurons, people, or packets of information. It is designed to be transposed from one research context to another as a tendentially neutral "model template" (Knuuttila and Loettgers 2014a, 2014b) that may be used to unearth the dynamic behaviour of a system of interconnected variables and hence imbue the target system – the brain – with complexity (Levin 2014).

Over the past decade, in particular, the discrete mathematics of networks has been naturalized as a description of the brain's internal dynamics via undirected graphs, which can tell you, for instance, that two neuronal populations are connected without providing any reliable information about causality (Sporns 2011). The focus now rests on interaction as the once-dominant idea of a well-ordered brain increasingly frays. Regions have turned into populations of neurons, and with finer resolution these populations might not appear to be homogeneous populations any more. The relational technology of the complex system is typically deployed "to understand things as purposefully conjoined but also to make and retool certain relations among particular known things" (Olson 2018, 6) (see figure 5.1).

What struck me, after my colleague had asked me about the significance of graph theory, was that Gareth had derived an analogy for the brain from a tool for data analysis by projecting the abstracted architecture of that tool onto the brain. While he was searching for

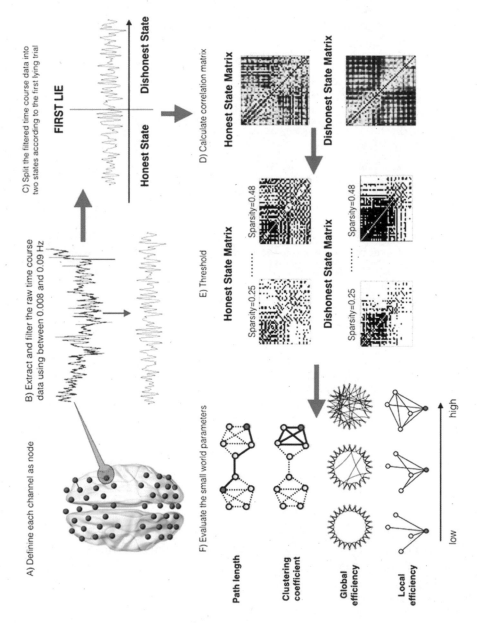

A) Definine each channel as node

B) Extract and filter the raw time course data using between 0.008 and 0.09 Hz

C) Split the filtered time course data into two states according to the first lying trial

FIRST LIE

Honest State   Dishonest State

D) Calculate correlation matrix

Honest State Matrix

Dishonest State Matrix

E) Threshold

Honest State Matrix

Sparsity=0.48 · · · · · · Sparsity=0.25

Dishonest State Matrix

Sparsity=0.48 · · · · · · Sparsity=0.25

F) Evaluate the small world parameters

Path length

Clustering coefficient

Global efficiency

Local efficiency

low   high

Figure 5.1 Steps in the analysis of functional brain networks based on graph theory.

the right words to explain what was so novel about the "graph theory treatment of the way [the brain] is organized," the tool itself provided the means to speak about the brain and suggested an analogy that did not seem to come about, well, naturally: "global IT [information technology]."[1]

Indeed, I found this way of thinking about the brain by resorting to its technical organization – patterns of interaction and flows of information – to be characteristic of a cognitive neuroscience that is in the process of abandoning one of its most significant forms of representation. No longer geared toward mapping a world that ends where the skull provides an enclosure, brain imaging now provides data of the workings of a cognitive infrastructure that might profitably be monitored and (reverse) engineered on an entirely different scale. Here, not only is the tool not epistemically neutral, but it also weaves a certain ontology into the brain's biophysical reality and contributes to the sinking of the brain's anatomy into an invisible background. The deterritorialization of the network is simultaneously a gradual infrastructuralization of the brain's "hardware."

INFRASTRUCTURALIZATION

The term "infrastructure" originates in French civil engineering discourses and was adopted into the English language in the late nineteenth century, where it came to be used to refer to the integration of parts into a complex, manageable whole (Carse 2017). In contrast to similar concepts such as network and system, infrastructure suggests a relationship of depth and hierarchy, which came to play an important role in distributing the institutional responsibility for preparatory groundwork before the installation of the "actual" technologies. After the Second World War, the notion of infrastructure became inclusive of more abstract entities that go far beyond the usual "bricks, mortar, pipes or wires" (Bowker et al. 2010, 97). Infrastructure now includes protocols, standards, memory, and generally everything that subtends, structures, and organizes daily life in its experiential background (Edwards 2003).

Infrastructure is typically neither invisible nor silent yet atmospheric in character as long as it does not make for a monument of cultural achievement or suffer from sudden breakdowns (Larkin 2013, 2016). Its disappearance from experience is a consequence of "infrastructuralization," such as in the case of more traditional infrastructure like highways and waterways, which tend to vanish from collective consciousness due to the fact that their existence is taken for granted and effectively "naturalized." The notion of the "infrastructural" denotes elements of a system or whole phenomenon, which we should not (have to) take care of.

In an article on the "ambient infrastructure" of generators in Nigeria, Brian Larkin (2016) describes how the initially sparsely distributed supplements to the national energy grid developed into a shadow system that, despite interrupting the original system of energy supply to a certain degree, depends on the national grid. That Nigeria's generator grid can be regarded as ambient might be surprising, for it is unbearably loud and typically wrapped in clouds of exhaust fumes. When Larkin uses the term "ambient," however, he refers also to the fact that the system of generators developed to back up the faulty national energy grid has meanwhile come to surround and interact with the national grid on a daily basis. If one system breaks down, the other jumps in.

The story about how the default network and its unconstrained, ambient activity came to reorganize our conception of brain function as defined by stimulus and response is surprisingly similar. The default mode, as a descendant of brain imaging's resting state or that which has been classified as noise, came into being only through analyses of what happens when the system that subtends external attention breaks down. Only gradually has it been decoupled from the resting state, yet it remains sutured to the empiricism that it came to interrupt in the first place. Intrinsic – "unconstrained," "self-generated," and "spontaneous" – brain activity increases when the demands of the environment abate, yet it does not, as researchers initially suspected, switch off in times when we are apparently focused on what happens in our surroundings.

The default network is considered to run in our experiential background and to support the externally activated brain in that it processes input according to its very own logic in order to adjust our cognitive models of the world. This idea implies that the default mode is relentlessly and, indeed, restlessly active but rarely stands out from the crowd of chattering neurons that surround it. It selectively links up to other large-scale networks in order to trigger specific cognitive "states," which are prominently involved in now highly valued cognitive capacities such as imagination and creativity.[2] The default network thus amounts to our most significant cognitive infrastructure in that it sustains individualized cognition yet can hardly be consciously controlled.

However, whereas the default mode has kept its status as the information processing system that is particularly active when we are slacking – not least due to its name, which binds it to the default mode of function that came to prominence through computing – the discovery of the myriad interactions that the default network has with other, task-specific cognitive systems has contributed to the notion that the brain as a whole might be more effectively analyzed as a highly complex cognitive infrastructure that operates according to the logics that the default network exposed. Identified through traces of activity in a purportedly resting brain and introduced as the ambient infrastructure of the creative, social, and intelligent subject, the default mode has inverted the concept of *mise en abyme*. Instead of being placed into the abyss, it emerged from an ocean of noise and now represents a decisive element of a renewed cognitive neuroscience that understands activated regions as hubs in distributed "cognitive systems," which interact with and modulate each other in response to cognitive demands. Mirroring the characteristics of the default network, the brain is in the process of being infrastructuralized; it now appears as an integrated piece of cognitive infrastructure that "emerges ... in practice, connected to activities and structures" (Bowker et al. 2010, 99).

As I suggested at the beginning of this section, the process of infrastructuralization typically involves making something sink into an invisible background in order, for instance, to conceal vital technology from the public eye and render the complexity of planetary-scale communication infrastructure manageable (Star and Ruhleder 1996). The abundance of projects that promote "infrastructural tourism" and that supply "field guides" to those seeking to reclaim the seemingly passive materials of an always-on society bears witness to the fact that what regulates our daily lives is often bundled, buried, and placed behind closed doors – or its presence has simply ceased to occur to us after decades and decades of routine use (see Burrington 2016; Mattern 2013; Mendelsohn and Chohlas-Wood 2011).

In the case of cognitive neuroscience, this aspect of infrastructuralization applies to what I have been calling the biophysical reality of the brain and involves the waning importance of cerebral geographies in favour of attempts to model network traffic within the brain. The exact meaning of infrastructuralization in this regard crops up most distinctively in comparison to another project that targets an infrastructuralization of the brain. The Human Brain Project's efforts to simulate the workings of the human brain on supercomputers, however, have entirely different vectors and have therefore been extensively criticized by those who advocate either computational approaches to brain function or, indeed, neuroscience-inspired artificial intelligence. The following section provides a brief overview of the HBP and analyzes how its plan to provide an infrastructure for the simulation of the human brain in silico now appears incompatible with the parallel process of the brain's infrastructuralization.

THE HUMAN BRAIN PROJECT

On 7 July 2014 members of the European neuroscience community published an open letter addressed to the European Commission concerning the Human Brain Project that criticizes its approach as "not neuroscientific enough" to further our understanding of the human brain (Abeles et al. 2014). In the letter, the signees state that the European Commission "must take a very careful look at both the science and the management of the HBP before it is renewed." At that time, the HBP was approaching its first decisive review phase with a tranche of 50 million euros in funding on the line.

The open letter expresses concern with the composition of the review panel, its evaluation criteria, the lack of an independent steering committee, and the uneven allocation of funding within the project. Many neuroscientists are further disturbed both by the project's megalomaniacal objective, namely initiator Henry Markram's vision to construct a neuroscientific complement to the Human Genome Project, and by co-executive director Richard Frackowiak's (2014) statement that the project is supposed to be "a CERN for the brain." Comparing the HBP with the infamous particle-physics laboratory near Geneva in Switzerland might not have been the most fortunate move, but the most prevalent concern of the critics is that the HBP has been presented as an attempt to understand the brain, siphoning funds from "real neuroscience" in the pursuit of computational infrastructure with no direct impact on understanding the brain.[3]

Slightly before July 2014, the HBP consortium's plans to dissolve its only dedicated cognitive neuroscience subproject had incited the resignation of eighteen involved laboratories from the HBP consortium, thus intensifying many neuroscientists' aversion to the project. Researchers at major European theoretical neuroscience institutes like the Gatsby Computational Neuroscience Unit in London, the École Normale Supérieure in Paris, and the Edmond & Lily Safra Center for Brain Sciences in Jerusalem had openly criticized the HBP for its purportedly misguided approach to studying the brain. In a comment published by *Nature*, Yves Frégnac and Gilles Laurent (2014) polemically ask, "Where is the brain in the Human Brain Project?" They paint a scenario where the Human Brain Project has explicitly eliminated neuroscience in favour of purely technological objectives. Time and time again, the project has been emphatically distinguished from proper, hypothesis-driven neuroscience as a project of exclusively technological value. A passage from Frégnac and Laurent's *Nature* article is exemplary in this regard:

> Neuroscience in the HBP is now limited mainly to simulations and to building a massive infrastructure to process mostly existing data. The revised plan advances a concept in which *in silico* experimentation becomes a "foundational methodology for understanding the brain." Numerical simulations and "big data" are essential in modern science, but they do not alone yield understanding. Building a massive database to feed simulations without corrective loops between hypotheses and experimental tests seems, at best, a waste of time and money. The HBP's goals now look like a costly expansion of the Blue Brain Project, without any further evidence that it can produce fundamental insights. (28)

The Blue Brain Project, now a subproject of the HBP, has been a significant touchstone for many critics. Launched in 2005 under the leadership of Henry Markram with funding from the Swiss National Science Foundation and the European Commission's Future and Emerging Technologies Flagship program, it completed its mission of simulating a neocortical column of a two-week-old rat brain in 2008. However, the first results were not published until 2015, when the Blue Brain project was already an integral component of the HBP (Markram et al. 2015). The effort was to no avail, for even after seven years of simulating an artificial cortical column, Markram was not able to convince his critics that the simulation was of any value in the search for the principles governing the rat's brain, let alone the much more complex human brain. In response to Markram's paper, Peter Latham of the Gatsby Computational Neuroscience Unit told the *Guardian* that the simulation "should really be thought of as a tool, something that can be used toward a deeper understanding of the brain" (quoted in Sample 2015), and should not be conceived as a comprehensive and game-changing model *of* the human brain.

Indeed, the reviewers of the HBP sided with the critics and recommended significant changes to management structures and governance as well as to the epistemological approach (European Commission 2015). In response to the recommendations, the HBP voted in favour of changing its governance structure and replaced the three-person executive committee with a director's board that consists of twenty-two members. Further, it reinstated the cognitive architectures subproject and launched a call for new projects to be integrated into its framework.[4] In a paper published in *Neuron*, the HBP's new scientific director, Katrin Amunts, and colleagues (2016) then redefined the project's objectives and explicitly emphasized the infrastructure aspect in the race to "decode the human brain."

It might seem as though the new executive, in the words of Éanna Kelly (2015), got the HBP "back on track" by downgrading a project that had targeted no less than reverse-engineering an entire human brain in silico. Despite the comparatively modest new ambition of developing infrastructure for the integration of data produced in neuroscience laboratories around the world, the critics remained suspicious of the HBP's plans and of the simulational approach in particular.[5] Kelly quotes Matthew Diamond of the Cognitive Science Sector at the International School of Advanced Studies, who maintains that the new agreement "can be, at best, an incremental adjustment of a strategy that is simply not right for life sciences, certainly not for neuroscience" – a problem that could have been circumvented, as Kelly was told by Guy Orban of the Department of Neuroscience at the University of Parma, if the HBP had been designed by neuroscientists instead of "artificial intelligence people or informatics people."

Orban's concerns about a possible alienation of the brain exhibit a surprisingly prevalent collective sentiment within the field of the neurosciences that certain questions should be reserved for researchers with "adequate" disciplinary backgrounds. Investigating human intelligence, creativity, and subjectivity, so the argument goes, must not be misunderstood as an endeavour of building more and more powerful computers and producing an abundance of data about the interaction of artificial neurons. On his blog *The Elusive Self,* cognitive neuroscientist Steve Fleming (2016) of University College London aptly remarks that expecting "big data approaches to succeed just because they have lots of data is like expecting to understand how Microsoft Word works by taking the back off your laptop and staring at the wiring."

What might at first glance look like Fleming's metaphorical way of explaining his aversion to an approach to modelling in silico must indeed be read as a critique of the HBP's tendency to reduce cognitive activity and hence the mind to an effect of its biophysical reality, or hardware. This is to say that Fleming's statement gets an entirely new twist when linked to what Jonah Lehrer (2008) had written eight years earlier about the HBP's predecessor, the Blue Brain Project, in a rather enthusiastic article for *Seed* magazine:

> [Co-director Felix] Schürmann leads me across the campus to a large room tucked away in the engineering school. The windows are hermetically sealed; the air is warm and heavy with dust. A lone Silicon Graphics supercomputer, about the size of a large armoire, hums loudly in the center of the room. Schürmann opens the back of the computer to reveal a tangle of wires and cables, the knotted guts of the machine. This computer doesn't simulate the brain, rather it translates the simulation into visual form. The vast data sets generated by the IBM supercomputer are rendered as short films, hallucinatory voyages into the deep spaces of the mind. Schürmann hands me a pair of 3-D glasses, dims the lights, and starts the digital projector. The music starts first, "The Blue Danube" by Strauss. The classical waltz is soon accompanied by the vivid image of an interneuron, its spindly limbs reaching through the air. The imaginary camera pans around the brain cell, revealing the subtle complexities of its form. "This is a random neuron plucked from the model," Schürmann says. He then hits a few keys and the screen begins to fill with thousands of colorful cells. After a few seconds, the colors start to pulse across the network, as the virtual ions pass from neuron to neuron. I'm watching the supercomputer think.

By focusing on the visualization of what the Blue Brain Project's supercomputer simulated, Lehrer provided an interesting vignette depicting why many critics of the HBP considered its simulations to be failures. The position from which Fleming and his HBP-critical peers now speak is not opposed to the fact that as part of the Human Brain Project the brain has found a new home in what could just as well be part of our planetary-scale cloud infrastructure. Nor is their critique necessarily directed against the simulation of biological (genetic), chemical (synaptic), and electric (neuronal) processes on arrays of massively parallel microprocessors that work according to the principles of computing *and* as a substitute for the complex biophysical system that sits between our ears. Instead, their critique revolves around the idea that the HBP approach will remain in the register of "*watching* the supercomputer think" (Lehrer 2008, emphasis added) and contribute to the idea that we can understand computations by "taking the back off your laptop and staring at the wiring" (Fleming 2016).

Although the extremely detailed simulations of the Human Brain Project, so the argument goes, might satisfy the desire for a god's view of the workings of an artificial brain, where you hit "a few keys and the screen begins to fill with thousands of colorful cells" (Lehrer 2008), recreating the brain in silico on a synaptic level must have appeared an unnecessarily complicated effort if the actual implementation of computation is, as Gareth told me, completely irrelevant to developing a coarse understanding of computational principles. In the comment section of neurofuture.eu, the very website where the open letter to the European Commission regarding the Human Brain Project was published, physicist Zoltán Toroczkai of Notre Dame University writes that the original HBP approach "of a bottom up simulation is akin to trying to understand the laws of gases by simulating the collisions of an Avogadro number (6.022×10^{23}) of particles" (quoted in Calhoun 2014), although these laws have already been derived from the phenomenological theory of thermodynamics and experimental evidence.

But there is more to the criticism than the impression that the HBP might constitute an unnecessary and costly effort. In reaction to the HBP strategy, Konrad Kording, then a professor of neuroscience at Northwestern University and a co-author of the paper discussed in chapter 1 that reports neuroscientists' inability to understand a microprocessor (Jonas and Kording 2017), listed a few examples of neuroscientific shortcomings in explaining the hardware of the brain on the website forum *Quora* (2014):

a) We know something about a small number of synapses but not how they interact
b) We know something about a small number of cell types, but not about the full multimodal statistics (genes, connectivity, physiology)
c) We know something about a small number of cell-cell connections, but a tiny fraction of all the existing ones
d) We know a few things about how a neuron's dynamics relate to its inputs, but only for a tiny number of cells and conditions.
e) We know a few aspects of a few neurons that change over time, but again for a tiny number of cells and conditions.

According to Kording, our limited understanding of the brain's workings on the level of its "hardware" calls into question the feasibility of attempts to simulate the brain down to the synaptic level, and as Fleming (2016) suggests, "[T]hings don't get any better if you have access to the 'firing' of every part of the circuit." Why re-engineer the human brain in all of its detail on supercomputers if the extremely limited insights into the biophysical reality of the brain would render any relevant progress in understanding the brain highly unlikely and all the more costly? The project is considered potentially damaging, for it might syphon funds from smaller, supposedly more promising neuroscience projects *and* because its epistemic surplus value is possibly limited to the field of computing and artificial intelligence. Hence it is the very "apparatus" (Barad 2007), an artificial reality of the brain in silico, that is considered inadequate to advance the production of knowledge about the workings of its biological siblings – not despite but because of its simulation of an entire human brain down to the level of synapses.

CLOUDS

The criticisms launched at the Human Brain Project and similar big data approaches within brain imaging in general exhibit the increasingly prevalent contention that investigating the biophysical implementation of consciousness might not lead to a more comprehensive understanding of what makes us creative and intelligent social individuals (Gomez-Marin et al. 2014). Indeed, neuroscience research shows signs of weakening its grip on the contrast-dependent, iconic brain of cerebral geography in favour of a more diversified, noisier, and more complex brain that is investigated primarily with regard to its information processing capacities or its technical protocols.

The emerging reality of the brain as cognitive infrastructure increasingly synchronizes the brain and what at present is the most significant metaphor of planetary-scale infrastructure and its organization: the Cloud. Indeed, much more than a mere infrastructure for the simulation of the brain's workings in silico (as in the case of the HBP), the Cloud simultaneously alludes to an ontological experiment that involves imagining the brain as a scaled-down version of cloud infrastructure. This emerging analogy between the brain and the Cloud involves thinking about cognition and intelligence in infrastructural terms.

Specific aspects that purportedly govern the placement and design of data centres, which form part of the Cloud, are also thought to characterize the cognitive architecture of the human brain, such as the minimization of wiring costs, the short path between nodes, and the dense local clustering of connections (e.g., see Menon 2011). The brain is thus reimagined as an ensemble of specialized small-world networks that are linked – just like data centre clusters in, for instance, the Netherlands and the United States – by a relatively small number of long-distance connections.

Similar to how contemporary cloud-rendering endeavours conceive data centre clusters, undersea cables, and cable landing zones as elements of rather abstract, planetary-scale networks of communication and information processing, cognitive neuroscientists have begun to understand brain structures as plastic, enabling resources instead of hardwired circuits (Bruder 2017; Mattern 2018; Starosielski 2015). The concept of circuits in the brain has therefore been reconceived: it denotes the recruitment of certain ensembles of neurons for specific cognitive functions, such as mental time travel and autobiographic memory (the default mode network), the selection of self-generated thoughts and externally generated stimuli for behavioural purposes (the salience network), or the transformation of processed experience into strategies in the pursuit of certain goals (the executive control network). Just as in the case of the Cloud, mapping and understanding the detailed wirings and mechanisms of packet switching remain extremely important for this endeavour, yet they are of minor importance in regard to understanding the interactions of large-scale circuits, as these interactions vary to a significant degree from subject to subject.

As I have tried to convey throughout this chapter, infrastructure is characteristically unspecific; when we talk about infrastructure, we rarely speak of *the* infrastructure. Even if we talk about global IT, for instance, we allude to communication infrastructure in general, not to undersea cable landing zones in the Azores or data centres in Iceland.[6] If you are using Microsoft's Azure Cloud Services, what Gareth called "the actual implementation," which comprises the geographical position of the servers, their hardware base, and the physical connections in between server farms, is essentially irrelevant as long as the computational task at hand is successfully completed. Infrastructure is typically simply infrastructure: something that we should not have to take care of as long as it works.

Cerebral geography, or the endeavour to map the brain's shapes in order to understand how it works as a system, will therefore likely suffer the same fate as cloud physiognomy. In "Cloud and Field" (2016), Shannon Mattern reminds her readers of a historical example of cloud iconoclasm, which arose when Johann Wolfgang von Goethe, inspired by the cloud studies of English chemist Luke Howard, tried to convince Caspar David Friedrich to produce a cloud atlas but was outright refused. Mattern quotes Friedrich as explaining "that 'to force the free and airy clouds into a rigid order and classification' would damage their expressive potential and even 'undermine the whole foundation of landscape painting.'"

Nevertheless, various volumes of the World Meteorological Organization's *International Cloud Atlas* were issued at regular intervals between 1896 and 1987, gradually forcing the capacious, flighty, and insubstantial figure of the cloud into a rigid system of classification. Lorraine Daston (2016) accordingly describes cloud atlases as an "achievement of collective seeing and naming [that] was made possible by terse descriptions that focused attention on a few key details and – even more important – obscured a myriad of others" (48). Early cloud observers, in order to contribute to the *International Cloud Atlas*, "had to learn to see the sky in the same way, to divide up the continuum of cloud forms at the same points, to connect the same words to the same things" (52). This is to say that the shape of clouds was also defined through templates and frames developed for their observation since the object in question was characterized most significantly by its ethereality.

Cloud physiognomy, or the fascination with *the shape* of clouds, came to an abrupt end with the advent of planetary-scale communication infrastructure. In *A Prehistory of the Cloud* (2015), Tung-Hui Hu observes that "engineers at least as early as 1970 used the symbol of a cloud to represent any unspecifiable or unpredictable network, whether telephone network or Internet" (x). He alludes to the example of AT&T's early videoconferencing system Picturephone, which operated regardless of the type of physical circuit underneath and was therefore depicted in diagrams as an amorphous form. In consequence, the Cloud is typically unbearably absent, being occluded by the puffy, celestial gestalt that provides planetary-scale communication and information processing infrastructure with a hypericon, suggesting that cloud infrastructure is an inexhaustible, limitless, and invisible reservoir of computing power and storage that can be operationalized at will.[7]

Just as obscuring the vastness of energy-hungry computing infrastructure enables our current model and usage of the Cloud, the renunciation of mapping every detail of the brain's hardware allows us to model and simulate its information processing capacities in a much more abstract way. The brain-as-cloud increasingly eludes static grids and spatial coordinates, re-emerging as a highly plastic cognitive infrastructure that is of characteristically uncertain shape. It points toward a dynamic understanding of materiality, which seems to defy attempts at detailed mappings and retrospectively discourages traditional, contrast-focused cerebral geographies. Against this background, the experiment conducted by Jonas and Kording (2017; see chapter 2, 21–4) can be interpreted as one instance in the gradual realization that, if the brain can be mapped only by means of partial templates and statistical techniques of averaging and normalization, an ever more detailed cerebral geography might never lead to a comprehensive understanding of how the brain does what it does – for the clouds of data that neuroscientists produce will ever remain in the midst of formation.

Instead, vagueness and volatility have increasingly come to define the brain-as-cloud; novel brain images, just like clouds in paintings, therefore figure as "harbingers of a new kind of image, an abstract one of flow and turbulence rather than symbolic representations" (Peters 2016, 58). It seems as though, at least at the computational end of contemporary cognitive neuroscience, pervasive visibility is about to be traded in for a much more blurred vision that develops *within* and is nurtured by the cloud of noisy data that the brain has gradually become.

In the computational paradigm of cognitive neuroscience, form is no longer imposed on the brain but "emerges," as Brian Larkin (2015) maintains for the case of infrastructure in general, "from the technical working of the object, which generates patterns that develop out of the operation of the infrastructure itself." Statistical devices from the domain of graph theory, for instance, provide the means to model the brain's information processing capacities via the co-activation of neuronal populations without alluding in detail to its "actual implementation" (Gareth).

As the brain now emerges as an entirely uncertain gestalt, it might not come as a big surprise that cognition is increasingly modelled on the experimental cognitive devices that have been designed to roam the clouds of data we produce through mediated interactions day in and day out. Such algorithmic "assemblages" (Rieder 2017, 113) extract patterns of information from social network data as well as from the signals that neuronal ensembles emit in scanners. They provide evolving models of cognition, whether in silico or in the brain, and do so largely in ignorance of the specific hardware base where the data are produced. The HBP has latched onto a recent surge in the development of cognitive computing solutions that build on systems assembled from graphics processing units (GPUs), and it has incorporated new prototypes for physical neural networks, but a computational version of cognitive neuroscience that developed in parallel to the HBP purposefully avoids deriving biological plausibility from biological fidelity.[8]

The infrastructuralization of the brain in computational cognitive neuroscience has involved a shift away from (hardware) circuits and toward code as well as away from questions that Louise Amoore (2018) regards as typical for "mode I" of a geography of cloud infrastructure – where is it? can we map it? – and toward an interest in processes and flows. Furthermore, it has involved a shift away from a particular history of observation and toward a bundle of experimental algorithmic techniques that "render perceptible and actionable that which would otherwise be beyond the threshold of human observation" (4).

Once more, the Cloud thus provides a fitting analogy. As Amoore (2018) observes, "in *cloud geography II*, where we are interested in the analytic, it is not so much the 'where' of the data that matters as the capacity to extract patterns in information, indifferent to the location or data type" (13, original emphasis). This situation applies to the Cloud, where the significance of the territory that houses data centres and network nodes is overridden by the organizational principles that undergird the system, as well as to the brain, where the specific location of activity is marginalized in favour of co-activations in a distributed network of neuronal populations. As the brain is increasingly conceived to be plastic in its functional organization, the question is not so much where activity occurs but how it can be modelled and predicted with sets of algorithms. Infrastructure is the interface of multiple, nested scales, and infrastructuralization is a means "of finding, within overlapping and at times unthinkable systems, point of view" (Lemenager 2016, 28).

The infrastructuralization of the brain through computational cognitive neuroscience proceeds within a framework of "calculative reason" (Carse 2017, 28) that stresses logistical and managerial perspectives on brain function instead of ever new mapping endeavours. Analyses and models of interactions between distributed populations of neurons, represented as clouds of data, characteristically hover above the territories of the brain: they refer to specific brain structures, yet it is the cloud of data itself, the "active perceptual system" (Haraway 1988, 583) where, for instance, task-specific networks are conceived. The network sits at the historical juncture of mapping and modelling and supports the emergence of object-oriented thinking in cognitive neuroscience: instead of tirelessly mapping the cerebral infrastructures of cognitive objects, ecological models of cognition are programmed to observe how the networks of cognitive objects unfold.

What I have described so far may thus be understood – by reference to a decisive attempt to rethink brain imaging experiments – as a "flipping of contrast" (Callard and Margulies 2011, 233). Whereas cerebral territories that ground cognition are gradually slipping from view, another aspect of infrastructure has come to the fore: the technical protocols or algorithms that govern how we make use of the brain as cognitive infrastructure.

6

Coding Cognition

In "Infrastructures as Ontological Experiments," Casper Bruun Jensen and Atsuro Morita (2015) envision infrastructures as a form of "practical ontology" (82), which involves "negotiations about standards" (82, citing Bowker and Star 1999, 44–5), a multiplication of "practical metaphysics" (82, quoting Latour 1999, 287), and "novel configurations of the world and its elements" (84). Infrastructures are experimental, the authors argue, since they open up a space of negotiation where concrete practices give rise to "objects that create the ground on which other objects operate" (82, quoting Larkin 2013, 329). The infrastructuralization of the brain is such an ontological experiment in that it scaffolds a novel cerebral alley where investigations of information processing independent of a substrate can take place. In this final chapter of the book, I focus on the effects of this ontological experiment, which conceives of the brain and the Cloud as analog, cognitive infrastructures on different scales.

The infrastructuralization of the brain is geared to rethinking human cognitive capacities in systematic, processual, and operational terms, yet instead of defining and stabilizing the brain, infrastructuralization introduces a productive uncertainty that defies mapping and multiplies ontologies. In her book *Into the Extreme: U.S. Environmental Systems and Politics beyond Earth* (2018), Valerie Olson observes that the concept of the complex system "is less useful when it is deployed only as a relational technology of enclosure; it is more powerful when it is worked with as a provisionally open-ended and aspirational form" (218). Form does in this case appear in its most abstract mode: not as shape or stable substrate but as formal organization and evolutionary trajectory.

At the heart of this reformulation sits a turn toward calculative and managerial techniques, such as simulation and prediction, that emphasize the generative dimensions of data analysis. Where experimentation on the lab bench is substituted with the simulation of algorithmic models, general purpose tools, such as graph theory, turn into hypothetical machines that bring forth new instances of how the brain and the Cloud might be related and reconceived. Yet, as this process proceeds, new programming practices and thus new stakeholders are homing in on the brain and helping to reimagine, by means of their very own epistemologies and techniques, cognition in the human brain and beyond. The Bayesian brain provides a specific example, where statistical techniques are translated into generative models that hover in between the life sciences and artificial intelligence research.

ARTIFICIAL REALITIES

John struck some keys, and a semi-transparent rendering of what looked like a grey rock filled the screen. Inside, a network of orange nodes and edges of varying thickness appeared and gradually started to dominate the enclosed space. Leaning back in his chair, John turned to me with a cheeky smile. Of course, I knew that I was looking at an experimental visualization of a brain "at rest" between two visual tasks, as we had discussed the experiment before. But I also knew John well enough to see that he was impatiently waiting for a surprised look on my face, and I definitely wanted to know more about the "glass brain," so I listened carefully to his explanation of the brain's "resting state" by other means:

> It seems that there is this consistent set of brain regions that
> are interacting strongly when people are relaxing on their back,
> not having much sensory input. And this is again something
> you can understand in terms of regions, but you understand it
> much better if you analyze the patterns of relationships, like the
> system type of view. So the first reason is that this is the way the
> brain works. And the second is that it's a new trend in science in
> general, the study of complex systems. Biologists are doing this
> and physicists have been doing this for a long time. It's basically
> this kind of large-scale organization of complex systems, which
> seems to be, based on mathematical graph theory, a good way of

explaining many phenomena in the world. So, you know, neuro-science is also catching up to this and maybe that's a good way to model the world in general.

I inadvertently smiled at the seeming circularity of John's account when he described the graph theory treatment of the brain both as a necessity since "this is the way the brain works" *and* as "a new trend in science in general." Yet, skimming through the transcripts of my recordings, I realized that his statement was not trivial after all: as strange as it might sound, brain mapping was not necessarily geared toward explaining how the brain "works." Whereas the epistemology of cerebral geography consisted of mapping the brain's functional regions, the use of graph theory in brain imaging data analysis involves a shift toward a focus on the time-based interaction of distinct neuronal populations. This shift, although disentangling neuroscience from geographical epistemologies, suggests novel analogies and opens up a space of interdisciplinary cooperation between those interested in understanding the brain and those primarily interested in the development of data models.

This is to say that John's work, like that of so many other methodologists, had a double bind: he was interested in elucidating the workings of the brain but only as far as graph theory and multivariate statistics, which project "the world" as a complex and uncertain system, could be improved through their application to brain imaging. Seeking to understand the world in general by means of certain statistical tools, my precious "outsiders" understood their work in the brain sciences as diverging from "real neuroscience," or the biologies that neuroscience supposedly holds onto.

The danger inherent in computational approaches – that analogies emerging through analytical tools might be "forced onto" the diverse research objects of complex systems theory – did not seem to matter so much. In the eyes of engineers, the molecular biologies of the brain represent but another abstraction that is considered outdated and "reductionist" exactly because it relies on specific structures and substances. As noted by another interviewee, Marta, a Spanish postdoctoral researcher working in a British brain imaging lab, bypassing the biophysical reality of the brain now figures as an approach that is complementary to a science that is conducted "within a traditional world of a lab where people are sitting at lab benches the whole day."

The lab reference calls up an array of science studies literature dedicated to the analysis of how computers have been transformed into the lab benches of a science 2.0, where computational methods figure as a form of experimentation in silico that differs more or less significantly from classic experimental paradigms and might therefore contribute to the emergence of novel research objects (e.g., see Keller 2003; Morgan 2004; Sismondo 1999, 2011; Winsberg 2010). Despite differences, all these approaches have in common that they focus on how computer simulations – in the tradition of the physical space of the "wet" laboratory – "reconfigure" objects and phenomena to render them fit for experimentation.

However, what being "fit for experimentation" means in "dry labs" might be very different from what it means in classic wet labs. The mathematician James Glimm, for instance, hypothesized that "scientific phenomena, once expressed in mathematical terms, could be solved numerically, without recourse to routine or repetitive experiment" (quoted in Gramelsberger 2011, 12). The overarching idea is that dry labs will relieve scientists from the burden of wet experiments, where they have to resort to idealized cases and model organisms that rarely live up to the complexity of the research object or that differ decisively from it.

In his essay "Computer Simulations and the Trading Zone" (1996), which he frames as a "workplace history" (120), Peter Galison accordingly describes computer simulation as "the artificial reality" (120) *of* a specific experimental setup, summarizing how the traditional professional categories of experimenter and theorist were challenged by electrical engineers and computer programmers in high-energy physics. Harking back to how distinct figures such as Stanislaw Ulam, Enrico Fermi, John von Neumann, and Nicholas Metropolis appropriated the concept of experiments for computational labour, Galison remarks that the early reality of high-energy physics "existed not on the bench, but in the vacuum-tube computers" (120) of the protagonists – who had "built an artificial world in which 'experiments' (their term) could take place" (120).

In particular, Galison (1996) refers to "a mode of inquiry" that was called "'Monte Carlo' after the gambling mecca" because "the method amounted to the use of random numbers (à la roulette)" (119). He explains that "physicists and engineers soon elevated the Monte Carlo above the lowly status of a mere numerical calculation scheme; it came to constitute an alternative reality – in some cases a preferred one – on which 'experimentation' could be conducted" (119). Here, the concept of "artificial reality" suggests that another, independent reality existed and defied experimental control. The example of the Electronic Numerical Integrator and Computer (ENIAC), however, offers another possible reading.

Indeed, in the early 1960s, Monte Carlo had not been employed to construct an artificial reality that was meant to simulate physical processes but had been conceived as a strategy to promote novel theories through the production of outputs that "looked like" experimental results (Borrelli, forthcoming). Only at the end of the 1960s did the epistemological status of Monte Carlo computations shift away from the simulation of experimental results and toward the simulation of experiments. What had started out as a mere cognitive tool in pursuit of solving putatively unsolvable integro-differential equations eventually merged the formalism of Monte Carlo into the then physical reality of the experimental setup, suggesting the idea of "computer-as-nature" (Galison 1996, 121).

What has been dubbed a switch from the bench to the screen might indeed be more productively described as a displacement of statistical techniques to the world of nuclear physics. Whereas simulation typically involves attempts to mimic and imitate some physical process, the example of ENIAC shows how the statistical devices of the Monte Carlo method initially provided the means to model particle resonance – a phenomenon that could not be observed in experiments and manifested itself only on the statistical level.

Computer simulations did not replace real-world experiments but instead introduced computation as a novel way of projecting a world that could not (yet) be observed. Statistical devices developed a new life in nuclear physics and helped to establish a third, "liminal" space where calculation schemes and empirical, datafied realities came to profit from and merged into each other.[1] Instead of providing a ready-made system for in silico experimentation, the ENIAC programs, and hence paradigms of contemporary computing in general, evolved *in exchange with* the phenomenon that they were designed to explore (De Mol, forthcoming; see also Haigh, Priestley, and Rope 2016). Instead of providing an artificial reality *to* the world of high-energy physics, the simulations conducted using ENIAC would hence have to be understood as creating various artificial realities that hovered in between the nuclear-physical and the computational.

Simulation effectively mediates between various real-world phenomena that might be productively compared not through reference to a closed, relatively stable, and independent reality but by assembling into "algorithmic techniques" (Rieder 2017) sets of rules that *might* determine their behaviour. Simulation accordingly figures as an ensemble of practices or "hypothetical systems" that destabilize the ontology of the phenomena it involves. As Tarja Knuuttila and Andrea Loettgers (2014b) observe, such "analogical reasoning provides modelers with a powerful cognitive strategy to transfer concepts, formal structures and methods from one discipline to another" (77). Yet it also entangles distinct research objects with an open-ended and highly tractable computational ecosystem that generates and conjures novel realities that hover above or in between, for instance, the brain sciences and a more general approach to cognition in silico.

FROM (HYPOTHETICAL) SYSTEMS TO (HYPOTHETICAL) MACHINES

In a bold move, experimental psychologist Bradley Love titled one of his articles "The Algorithmic Level Is the Bridge between Computation and Brain" (2015). His argument refers to David Marr's infamous trilevel hypothesis, which suggests that information processing has to be understood at three distinct yet complementary levels of analysis: implementation (hardware), algorithm (program), and computation (program specification or abstract problem description). However, Love conceives of this tripartition as an unnecessary complication of the matter, one that might indeed hamper the neuroscientific project.

He suggests bridging the three levels of analysis by limiting investigations to the algorithmic level. In other words, Love's initiative is not geared toward comparing the brain and the computer – as in the tired brain-computer metaphor – but seeks to establish a common ground on which information processing in humans and machines may be mutually remodelled.

Love (2015) opts for the algorithmic level as purportedly representing the least biased level of analysis since it is determined neither by the observed hardware activity (implementation) nor by the specific (biological) substrate. Nevertheless, the algorithmic level may be "used to explain implementation-level findings, including brain imaging results, as well as to link to formalisms that are optimal in some sense (i.e., reside at the computational level)" (231–2). In short, biological fidelity is not a requirement but a welcome surplus.

Investigating the algorithmic level of brain function therefore constitutes a rather intuitive and playful task, where biology fades in and out of view but typically vanishes behind the interface of programming environments and engineering tools such as MATLAB, discussed in chapter 2. One of my interveriewees, Lance, a post-doctoral researcher at a Swiss neuroscience institute, told me in an e-mail that he did most of his work in front of a computer, "playing around with MATLAB." When I asked him what exactly that meant, he explained,

> Ah, MATLAB is a language that's very abstract, so it's not like a programming language like C++ or C, where you really have to think everything through before you go and try different things. With MATLAB, "playing around" basically means trying different ideas that you might have, kind of quickly with the data, and trying different ways of representing the data and different ways of making a model of the data so that you can see what changes in the result ... because for me, if I would have to do maths, I would do it probably on paper or take a notebook etcetera. But then once I have a rough idea of what the problem looks like, then playing around in MATLAB means really kind of loading the data and visualizing the data and trying to extract some features from the data and make a model of it and try to make some predictions. But to me, it's really like embodying your intuition as an algorithm but not as a single algorithm, as a series of algorithms.

Modelling brain imaging data, as Lance describes it, is a characteristically iterative process that proceeds largely in the absence of volunteers in scanners. Rather than simply reproducing individual experimental results, the model is programmed to live up to the noisy reality that derives from, for instance, temporal fluctuations in the network interactions of neuronal populations. As regards testing models, noise is not necessarily considered an issue; in the pursuit to program more robust models, it apparently makes for a virtue.

Models in MATLAB are built on the empirical ground truth of brain imaging data. The prototypes are thus first simulated and used to produce artificial data sets in order to prove that the prototype is not "buggy." Said procedure is typically called "inversion." As noted by another interviewee, William, the head of methods at a British brain imaging lab, it helps researchers to gain an understanding of "the behaviour" or "an intuition for how it works." Inversion involves the playful tweaking of parameters within the model to understand the variance it generates. John stated, "It's like an iterative improvement because you have a supposition. Then you try it in MATLAB quickly, play around, and see the result, and it's like you thought or not."

In this process, mathematical formalism is moulded in response both to the specific data set and to the requirements of simulation in MATLAB. The oft-reclaimed mathematical certainty and precision are to a certain degree undermined by an intuitive approach to assembling various statistical techniques and mathematical theorems into algorithmic techniques, or into a "logical series of steps for organizing and acting on a body of data to quickly achieve a desired outcome" (Gillespie 2016, 19). The noisy empirical reality emerging from activity of the brain in the scanner provides the hypothetical system with a reassuring ground truth and allows for a playful engagement with abstract mathematics. MATLAB in neuroscience puts theory in motion.

As Tarleton Gillespie (2016) puts it, algorithms determine a procedure "in the service of the model's understanding of the data and what they represent" and in the service of "the model's goal and how it has been formalized" (20). This is to say that data-driven models of brain dynamics are not only judged with respect to the numerical data they produce but also figure as descriptions and ideal explanations of how a set of data might have come into being. The sort of data models that methodologists compile in MATLAB turn "hypothetical systems" into "hypothetical machines" in that they exhibit certain general features of the phenomena they are supposed to explain and simultaneously variegate the mathematical tools or algorithmic techniques they involve.

Algorithmic models increase the general tendency of models to become "multiplex and unfolding" (Merz 1999) entities that elude a specific purpose, representing instead instantiations of an ongoing process of modelling novel ontologies. The resulting models are in a state of becoming and are "teased into being: edited, revised, deleted and restarted, shared with others, passing through multiple iterations stretched out over time and space" (Kitchin 2017, 18). As a result, they are always and deliberately uncertain and provisional entities that facilitate the nearly seamless transition away from traditional cerebral geographies – via what I have been calling geographies of information characteristic of network neuroscience – and toward algorithmic views of brain function. This shift diverts attention to, and makes use of, for instance, fluctuations in neural network activity as decisive mechanisms of information processing (e.g., see Deco et al. 2017; Shine et al. 2016; Taghia et al. 2018).

Within this trajectory, models of information processing have evolved from programmed hypothetical systems to generative hypothetical machines that continue to produce novel readings and drafts of cognitive realities. The question of how – that is, in which programming environments and languages – such cognitive entities are coded is hence much more important than it might seem at first sight.

PYTHONIZED

In her book *If ... Then: Algorithmic Power and Politics* (2018), media researcher Taina Bucher rightfully argues that computation is independent of programming language and the specific programming practices used by coders. Indeed, computational epistemologies are not necessarily inherent to a specific programming language but are designed for and tend to lend themselves to specific tasks or approaches to computation. Accordingly, the introduction of novel programming languages will not necessarily change the shape of computed phenomena, but the case of object-oriented programming shows that certain techniques bring about novel "epistemologies," which might also have effects on the ontologies of objects outside of the computational realm. The gradual adaptation of Python as a master programming language in cognitive neuroscience promises to be another case in point, for Python comes naturally to younger programmers whose informational labour will in the long run also have an effect on models of information processing in human brains and intelligent machines.

Python is a much more general and significantly more flexible programming language than MATLAB and speaks to a novel breed of programmers who learn coding in a different way. Meanwhile, Python has been augmented with dedicated packages that target various disciplines, including Scikit-learn for machine learning,

Biopython for bioinformatics, PsychoPy for psychology and neuro-science, and Astropy for astronomers. Whereas MATLAB has been developed to open mathematical formalism up for coding exercises, Python is first and foremost an intuitive and highly readable pro-gramming language that provides a novel platform for integrating a large number of widely used simulation environments. The Python ecosystem allows researchers to combine elements of "different pro-gramming languages, which together facilitate the interaction of modular components and their composition into larger systems" (Muller et al. 2015, 1).[2]

Easy to read and accessible as a programming language, Python appeals particularly to those who excel in programming rather than in the sort of "playing around" (Lance) with mathematical formal-ism typical of the MATLAB ecosystem. Attendant is a shift from the classic, intellectual work of developing mathematical models to the much more physical epistemology of coding, which prominently involves the painstaking optimization and combination of code chunks or "reusable components" by hand.[3]

As so very often happens, recent developments in biology provide an instructive example of what the field of cognitive neuroscience might soon undergo. In an article published by *Wired* magazine, Emily Dreyfuss (2017) investigates the significance of coding skills for those who "want to make it as a biologist" and quotes a postdoc-toral student in genetics at Harvard, who explains that when he first watched an expert in computational modelling, he "realized that just the way he held his laptop was completely different from me. His fingers were spread wide open over the keys in this diagonal format, and I just knew I'm fucked, I'm fucked in this whole field" (see also Muller et al. 2015, 1).

The introduction of Python has most likely privileged a certain class of programmers, or "real coders," who are increasingly unfa-miliar with the traditional epistemologies of the life sciences and will work toward the ends of improving an evolving computational ecosystem. This process will continue to go hand in hand with the adoption of coding aesthetics that emerge through the use of Python. As suggested by Tim Peters's "The Zen of Python" (2004) – a collec-tion of guiding principles for Python programming communicated through nineteen aphorisms – the aesthetics of coding might grad-ually replace the aesthetics of the mathematical formalism that has come to undergird contemporary cognitive neuroscience:

Beautiful is better than ugly.
Explicit is better than implicit.
Simple is better than complex.
Complex is better than complicated.
Flat is better than nested.
Sparse is better than dense.
Readability counts.
Special cases aren't special enough to break the rules.
Although practicality beats purity.
Errors should never pass silently.
Unless explicitly silenced.
In the face of ambiguity, refuse the temptation to guess.
There should be one – and preferably only one –
    obvious way to do it.
Although that way may not be obvious at first unless
    you're Dutch.
Now is better than never.
Although never is often better than *right* now.
If the implementation is hard to explain, it's a bad idea.
If the implementation is easy to explain, it may be a good idea.
Namespaces are one honking great idea – let's do more of those!

Although coding aesthetics are certainly not all there is, they considerably influence computational epistemologies and modelling practices in the neurosciences through the material-semiotic practices of coders.[4] One aspect of "The Zen of Python" sticks out in this regard: "In the face of ambiguity, refuse the temptation to guess."

So far, the introduction of Python to the neurosciences has gone hand in hand with a turn toward machine learning in signal processing and data analysis, which has involved a recontextualization of traditional statistical techniques. As Adrian Mackenzie (2017) explains, contemporary machine learning is best understood as an accumulation of mathematical constructs drawn from linear algebra, differential calculus, numerical optimization, and probability theory, which "take shape against a background of more than a century of work in mathematics, statistics, and computer science as well as disparate scientific fields ranging from anthropology to zoology" (5). However, the action of reassembling various mathematical techniques proceeds largely independent of the contexts where they once emerged. Mackenzie therefore calls both programmers *and* algorithms "machine learners," for the former provide statistical techniques with a new life in the Cloud and the latter help us to relearn the world through a radical, data-based empiricism.

The logic of this learning process is based on prediction. In his book *Machine Learners: Archaeology of a Data Practice* (2017), Mackenzie states that artificial neural networks "pivot around ways of transforming, constructing or imposing some kind of shape on data and using the shape to discover, decide, classify, rank, cluster, recommend, label or predict what is happening or what will happen" (432). A comprehensive theory of real-world phenomena is hence ever postponed; at the heart of machine learning in cognitive neuroscience sits a computational epistemology privileging algorithms that "provide a minimal functioning starting point for exploring what biological details matter to brain computation" (Kietzmann, McClure, and Kriegeskorte 2019, 12).

Once inspired by the brain, highly plastic and continuously morphing artificial neural networks now project their organizational logics back onto the brain. The role of methodologists in brain imaging might thus continue to shift away from that of "ghostbusters" who program algorithms that ideally reproduce the empirical reality of brain imaging data sets and toward that of facilitators who code and recode free assemblages of various statistical devices that act as ghostly placeholders of an emerging, post-anthropocentric cognition in between neurobiology and AI.[5]

BAYES

In the introduction to this book, I quoted Demis Hassabis comment in *Nature* on the occasion of Alan Turing's centennial: "To advance AI, we need to better understand the brain's workings at the algorithmic level – the representations and processes that the brain uses to portray the world around us" (Hassabis, in Brooks et al. 2012, 463). In practice, however, modelling typically proceeds in the opposite direction. A group of neuroscientists loosely affiliated with Google's AI divisions, for instance, argues "that a range of implementations of credit assignment through multiple layers of neurons are compatible with our current knowledge of neural circuitry, and that the brain's specialized systems can be interpreted as enabling efficient optimization for specific problem classes. Such a heterogeneously optimized system, enabled by a series of interacting cost functions, serves to make learning data-efficient and precisely targeted to the needs of the organism" (Marblestone, Wayne, and Kording 2016, 1).

This description obviously does not seem to come naturally to (our understanding of) the brain; in order to provide an algorithmic description of its workings, the authors suggest using a statistical device that might gradually reproduce a noisy ground truth that consists of "real-world" or brain imaging data. Initially programmed for data analysis, machine learning algorithms never cease to optimize their cognitive capacities and integrate what they "perceive" into their cognitive architecture. What happens in this process is that the model changes from one that is based on the real-world data of neural network interactions to one that is largely an independent "organism" used to predict "what a brain *should* in fact compute for an animal to behave optimally" (Kriegeskorte and Douglas 2018, 13, emphasis added).

The very characteristics of mathematical formalisms implemented in machine learning algorithms translate into current understandings of brain function. Computational approaches to cognitive neuroscience are currently often linked to Bayesian models of cognition, which, as the name suggests, derive from Bayesian inference or, indeed, the statistical cultivation of doubt. Media studies scholar Bernhard Rieder (2017) accordingly analyzes the "'naive' Bayes classifier" (108) as an interested reading of a "datafied reality" (110) that projects "latent structures present in the data in a myriad of ways" (111) without ever accepting the data as depicting "reality." Such algorithmic techniques entail, to use Rieder's words, "a way of both looking at and acting in and on the world" (109).

An anecdote retold by brain imaging eminence Karl Friston shows how closely the idea of the Bayesian brain and deep learning might be related. In his article "The History of the Future of the Bayesian Brain" (2012), Friston sets out to create an autobiographical narrative from personal accounts and anecdotes, which, as he states in the introduction, may or may not be true. Geoffrey Hinton, a forerunner of the deep learning approach, appears in a later section titled "The Bayesian Paradigm." Friston recalls that Hinton, who had come to London to set up the Gatsby Computational Neuroscience Unit, told him about a new approach to unsupervised learning just before returning to Canada, which prompted Friston to formulate his ideas about the Bayesian brain and the minimization of free energy in biological systems (Friston 2010a, 2010b).

Whether the story is true or not, Friston's anecdote captures how deeply neural network models, the statistical principles of Bayesian inference, and emerging understandings of the human brain are now often entangled.[6] At the computational end of cognitive neuroscience, the statistical device *is* the hypothesis, turning into a hypothetical machine as it produces instances or "simulacra" of cognitive devices, which clearly belong neither in the domain of neurobiology nor in that of machine intelligence. This is to say that the Bayesian brain is not merely an ideal model of brain function; rather, by means of "deep belief nets" (Hinton, Osindero, and Teh 2006), which kicked off the current deep learning frenzy, algorithmic assemblages are also simulated as possible realities of cognition in between the poles of human and machine. The issue at hand is thus not whether AI can be considered more or less intelligent than its human others but whether the "human" intelligence that we know through psychology and the social sciences is no longer the uncontested benchmark.

No longer primarily an organ of the body defined by a biophysical reality or human behaviour, the human brain now appears as an organ of contemporary cognitive capitalism. Bayesian thinking, for instance, is at present very prominent in the design and conceptualization of smart cities, which mark "a turn to inductive reasoning, subjective perspective (there are no stable truth claims), and the abandonment of stable baselines or norms, a turn that finds itself incarnated in such ideas as 'data-driven' science, marketing, and strategy" (Halpern et al. 2013, 294). The Bayesian approach to brain function now projects these ideas onto the brain – a "Bayesian brain" (e.g., see Doya et al. 2006) that is of characteristically uncertain shape and defined primarily through capacities that seem to be entirely foreign to the brain that we have come to know through traditional neurosciences. Is it "the privileged machine in this context that creates its marginalized human others?" (Suchman 2007, 270).

Infrastructural Reveries and Psychic Lives

In one of the most significant papers of the history of deep learning, AI researchers Geoffrey Hinton, Simon Osindero, and Yee-Whye Teh (2006) argue that a goal of AI research with respect to an artificial neural network must be to metaphorically "look into its mind" (1529). Indeed, visualizing the detailed cognitive architectures of deep neural networks has so far played no more than a minor role in artificial intelligence research. Only very recently has Graphcore, a semiconductor company that develops accelerators for machine learning, broken the mould and published what it calls "AI brain scans" (see figures D.1 and D.2). Paradoxically, these images, although aesthetically valuable, illustrate the demise of the visual in computational cognitive neuroscience and AI rather than encouraging a return of "the image." Graphcore's AI brain scans might be reminiscent of cell microscopy, yet what they render are the statistics of a learning process that is independent of substance or structure.

AI brain scans may be considered a reaction to the ubiquity of discourses that revolve around the problematic black-boxing of (corporate) algorithms rather than an attempt to "look into" the mind of neural networks.[1] In fact, Graphcore's visualizations have to be understood as promoting a rather iconoclastic empiricism "that disperses patterns as the visible form of difference into a less visible but highly operational space" (Mackenzie 2017, 149).

AI brain scans promise to elucidate how an algorithm "makes sense" of what it "perceives," but what Graphcore anticipates and addresses are the various ways by which people try to "make sense of the operational logic that experts routinely describe as *unknowable*" (Bucher 2016, 82, original emphasis). This use of AI brain

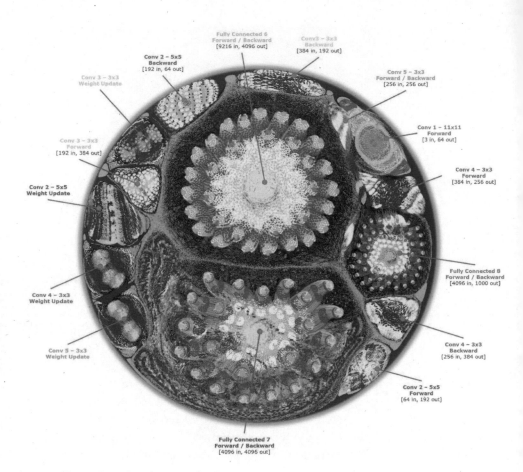

Figure D.1 Graphcore AI brain scan.

scans can be considered another parallel between contemporary AI research and cognitive neuroscience, where, as I have argued over the course of this book, X-ray-like brain images that show only statistically identified spikes of neuronal activity are in the process of being replaced by colourful visualizations of distributed network traffic in the brain. These visualizations no longer speak to the traditional cerebral geographies that were intended to map a stable and intersubjectively generalizable brain. Instead, they exhibit how brain imaging technology is repurposed within a novel, computational epistemology that emphasizes simulation and prediction in pursuit of an algorithmic understanding of brain function.

At its computational end, cognitive neuroscience is no longer a science that flirts with phrenological ideas and threatens, according

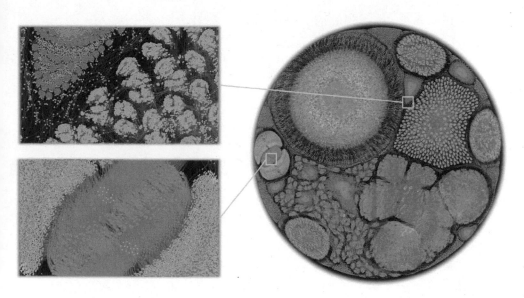

Figure D.2 Graphcore AI brain scan, detail.

to concerned psychologists, humanities scholars, and social scientists, to reduce the (social) subject to its cerebral substrate; instead, computational cognitive neuroscience delineates a field of research that appears to be drifting toward a liminal space of curiosity-driven exploration, where (social) cognition, intelligence, and creativity are about to be investigated by means of generative algorithms.

Somewhat ironically, the novel prominence of computational methods in cognitive neuroscience has its roots in the failure of brain imaging statistics. The discovery of data dredging practices in social neuroscience cast shadows of doubt over the existence of large, specialized brain regions, which have since increased due to the gradual reanalysis of data that was discarded as noise in accordance with traditional brain imaging methodology. This process brought a shadow system within the brain to the fore, which has contributed to changing our conception of the brain from the inside. The title of resting state research frontrunner Marcus Raichle's article "The Restless Brain: How Intrinsic Activity Organizes Brain Function" (2015) seems to capture how the idea that the so-called default network or mode is "the mind, which does not rest even when explicitly asked to do so" (Marchetti et al. 2015, 1), has contributed to thinking about the entire brain differently – that is, as the industrious cognitive infrastructure of a relentlessly information processing, human individual.

What has since emerged is an increasingly complex, restlessly active, and infinitely plastic brain that can be understood only with recourse to the temporal dynamics of neural activity. Brains are still being mapped – if only to discern the elements of their global communication infrastructure – yet the interest in longer and more frequent scanning runs conducted with volunteers devoid of cognitive input and greater spatial resolution suggests that the sort of task-based imaging that resulted in static maps with clearly delineated functional brain regions is gradually falling out of fashion.

Autism spectrum disorder, for instance, now appears most significantly by means of neuroscience techniques and technologies "that can be used while the participant is 'at rest or asleep' [and that] provide the opportunity to 'know what's behind' verbal behaviour and 'see the real, the true abilities of autism' in a way that is not dependent upon such abilities to perform" (Hollin and Pilnick 2015, 283; see also Hollin 2017). Instead of locating the spectrum within the hardwired structures of the brain by observing brain responses to "social stimuli" (Hollin and Pilnick 2015, 281), researchers now rely on the absence of cognitive input during brain imaging experiments, which reveals a much more spectral, much more complex, much more technical, and much more infrastructural condition.

The infrastructuralization of the brain involves a marginalization of performing bodies in the scanner room. Resting state fMRI provides images of the autistic brain that can be stared at from a position that purportedly lies outside the realm of experimental social psychology and psychoanalysis, reconceptualizing autism spectrum disorder as a variation in the "normal" patterns of co-activation of the brain's default network and brain structures responsible for modulating attention and cognitive control (e.g., see Padmanabhan et al. 2017; Singh 2015; Yerys et al. 2015).

In response to these developments, as Hallam Stevens (2013) observes in the similar case of bioinformatics labs, the informational labour of those who have turned to neuroscience "to find new kinds of problems to solve and new applications for their programming and algorithm-building skills" (51) is becoming ever more significant. Instead of being put off by the messiness of neurobiology, methodologists in brain imaging laboratories tend to embrace the uncertainty that derives from a steadily growing – or better, thickening – cloud of noisy data. The noisier the data, the greater the challenge to analysis models that are not necessarily brain-specific and may be employed for the modelling of seemingly very different yet comparably complex "real-world" systems.

Meanwhile, traditional techniques of neuroscience are far from useless. "They're effective readouts of health and illness, marking changes related to disease, learning, pharmaceuticals, and so on," neuroscientist Kelly Clancy argues. "But using them to sieve meaning about the fundamental logic of our nervous system is another matter" (quoted in Yong 2016). Instead, machine learning algorithms act as evolving hypotheses of how brains process information. It is considered a virtue rather than an issue that neural network algorithms, despite generating myriad connections among their multiple layers, are far from being able to reproduce the complexity of the human brain's cognitive architecture. "Who is to say that every feature of our brain is worth mimicking," Clancy (2017) asks in a column for *The New Yorker*, pointing to an iconoclasm that is gradually turning into a defining characteristic of cognitive neuroscience.

Graphcore's AI brain scans should accordingly not be mistaken for an application of traditional neuroscience techniques to cutting-edge AI. As shown by Eric Jonas and Konrad Kording's (2017) failed attempt to understand a microprocessor by imaging its hardware, the interest in hardware activity has waned both in the brain sciences and in computing. Instead, Graphcore's AI brain scans communicate the novel intimacy between neuroscience and AI by representing the algorithm as a quasi-biological organism that learns and evolves in analogy to the algorithms that *could* potentially govern the development of cognitive architectures in the human brain. The visual reference to cell microscopy promotes a revised understanding of the relationship between biology and technology, where biology turns into a mere infrastructural variable within a relentlessly optimized, cognitive system.

Analyzing Google's attempt to establish a liminal space between (neuro)psychology and AI, I was reminded of neurophysiologist Warren McCulloch's attempts in the 1940s to think through an experimental epistemology by means of "existential devices," or "theoretically conceivable nets" (Abraham 2003, 425), that made use of existing mechanisms to posit a possible material basis for a psychic structure, thus generating answers to the question about what we would experience if we were caught in this or that particular circuit.

The specific materiality of these circuits, their hardwired connections, and even their constituting mechanisms were imagined to be subordinate to how the circuits' behaviour changed over time. Or as McCulloch put it in 1949, "In the neurotic brain you may find no general chemical reaction gone astray, nor any damaged cells, for when activity ceases, regeneration ceases. The most you might expect to find are some changed thresholds or connections" (quoted in Dumit 2016, 231). Joe Dumit writes that McCulloch's existential devices

> were uncanny, weird, and irrational. Yet they were clearly logical. One could follow them along as they followed their instructions, step by step. And yet the result was often not rational. They would get caught up in a loop and have no way out, akin to people getting stuck in a rut or becoming obsessive. They would ignore everything else except their programming, like subjects of ideology. Little errors would compound into big ones. They had no way of reflecting on whether they were running the right program. In other words, different running computers were great analogs for types of personalities. They were logical [and] each in their own way, "patho-logical." (223)

Similarly, in neuroscience-inspired AI, pathology is no longer considered to be a result of immutable biophysical realities or persistent *traits* but first and foremost to be a bug in the system or a pathological *state* that might prove to be beneficial to the development of the organism or be open to attempts at recoding to prevent that the evolutionary trajectory "enter[s] a zone of qualitative indeterminacy" (René Thom, quoted in Bates 2014, 42).[2]

In mathematician René Thom's approach to modelling morphogenesis and structural stability as well as in epistemologist Georges Canguilhem's 1947 lecture "Machine and Organism," the organism remains tied to a biological substrate. Canguilhem (2008) differentiates between these two concepts by ascribing the "the rational norms of identity, consistency, and predictability" to the machine (90). Whereas the machine will stick to its intrinsic organization even in catastrophic events, an organism is defined by its polyvalent organs, which allow for a pathological response: it is "not simply the negative inverse of a normal function, as cybernetics at times seemed to imply, but rather an opportunity for the invention of radically new behavior" (Bates 2014, 35).

Canguilhem's idea of organismic identity had been considerably influenced by the deliberations of neurologist Kurt Goldstein, about whom Matteo Pasquinelli (2015) writes that he conceived the human brain as able "to invent, modify and destroy its own norms, internal and external habits, rules and behaviors, in order to adapt better to its own *Umwelt* (or surrounding environment), particularly in cases of illness and traumatic incidents and in those conditions that challenge the survival and unity of the organism" (83). In early cybernetic endeavours, the differentiation between machine and organism was performatively undermined by conceiving the brain as an ensemble of systems or circuits that regulate each other. Apart from McCulloch, John von Neumann and Norbert Wiener projected and experimented with machines that could potentially imitate the organism's ability to preserve unity at the expense of a particular internal configuration. Von Neumann (1966), for instance, elaborated on the impossibility of locating anything in the brain given its "enormous ability to reorganize" (49). Wiener (1948) found it hard to believe that a memory system could be stable for a long period of time since the unpredictable character of its environment suggests that it is ever on the cusp of information overload. The ever-looming, "qualitative indeterminacy" of an organism's evolutionary trajectory hence turned into the potentially catastrophic failure of a machine that is never allowed to slack.

POST-ANTHROPOCENTRIC INTELLIGENCE

In the design of Google's neuroscience-inspired gaming algorithm AlphaGo, Demis Hassabis and his fellow DeepMind engineers provide a sketch of how organisms – whether biological or in silico – *could* learn. Central to the modelling process is the idea that the human Go player no longer provides an ideal but acts as an adversary that trains AlphaGo to develop superhuman abilities. Whereas the first versions of the algorithm did have access to a vast database of possible moves, AlphaGo Zero learned to play the game from scratch. Devoid of the concept of fatal errors, the algorithm makes seemingly catastrophic decisions that play out in the end, which contributes to a sort of ecological "awareness" on the part of the algorithm that permits it to "simultaneously cope with the vast complexity of the game, the long-term effects of moves, and the importance of the strategic values of positions" (Gelly et al. 2012, 107).

Thanks to the combination of two artificial neural networks, one of which is responsible for calculating possible moves situationally while the other continuously revalidates long-term effects by comparing the current game with past experience, AlphaGo has developed a "feeling" for situatedness, or what in humans is referred to as "intuition." The title of Cade Metz's article in *Wired* magazine on the Google DeepMind challenge, "The Sadness and Beauty of Watching Google's AI Play Go" (2016), conveys the generally ambivalent sentiment toward the fact that a hopelessly rule-bound assemblage of rare earths, plastics, and coding skills suddenly looked like an unpredictable organism that even appeared to have a sense of aesthetics.

AlphaGo created the desire among professionals to play Go like the algorithm. Fan Hui, who had trained against the algorithm prior to the DeepMind challenge and had been devastatingly defeated in a closed-door match with AlphaGo a few months before, reportedly said of one of AlphaGo's moves, "It's not a human move. I've never seen a human play this move. So beautiful" (quoted in Metz 2016b). Just by training against AlphaGo, he had learned to see the board in an entirely different way and consequently saw his world ranking skyrocket. Hassabis himself reported that AlphaGo's second victim, the much more prominent Korean professional Lee Sedol, echoed the words of Fan Hui: "Just these few matches with AlphaGo, the Korean told Hassabis, have opened his eyes" (Metz 2016a).

Experts were apparently not impressed by the algorithm's anthropomorphic gameplay but by its ability to find novel strategies that had not occurred to any human player before. The algorithm was therefore hailed not despite but because of its exhibition of post-anthropomorphic strategies in a decidedly anthropocentric game. Human players can profit from watching the algorithm play and, by internalizing how it decides, augment their human thinking with machinic tenacity.

Can what is currently an algorithmic technique of resilient information processing also provide insights into how the industrious subjects of contemporary capitalism may be stabilized in order to prevent catastrophic failure in the face of information overload?

Clemency Burton-Hill in a *Guardian* article about Demis Hassabis titled "The Superhero of Artificial Intelligence" (2016) constructs a short-lived narrative arc that aligns creator and creature, human and machine by way of their work habits. In their conversations, Hassabis admitted that he tends to stay up till three or four o'clock in the morning and use the final, darker hours for unconstrained thinking. That is the time when his spouse and kids are in bed, Skype calls to the United States have been made, and cognitive input is rare. It is the time for the "default mode." Recounting one conversation with Hassabis, Burton-Hill writes,

> I'm reminded of AlphaGo, up there in Google's unimaginably powerful computing cloud, just playing and playing and playing, self-improving every single second of every single day because the only way it can learn is to keep going …
> "Does it ever get to rest?" I ask.
> "Nope. No rest! It didn't even have Christmas off."
> I hesitate. "Doesn't it ever need a break?"
> "Maybe it likes it," he shoots back, a twinkle in his eye.
> Point taken.

Not only in this article are DeepMind algorithms portrayed as hardworking, industrious organisms that never slack, but the game at which they excel also provides a cultural context that reinforces said impression:

> Legend has it that Go's invention came via two immortals who instructed the mythical king Yao to use it to teach his rakish son the merits of patience and calm reason. The chessboard represents an ancient view of the universe: Round stones representing heaven, a square board for earth, *yin* (black) and *yang* (white) forces acting out the processes of life amid the intersections that symbolize the days of the year …
> In the Shuyiji (述异记, *Tale of Strange Matters*), a fifth-century collection of mythical stories, there's the tale of Wang Zhi, a woodcutter who came across a group of children playing Go in the mountains. Wang sat down to watch for some time, until the children asked, "Isn't it time you went home?" He then stood up and saw that the wooden handle of his axe had rotted away. Returning to his village, everything looked different; local legend spoke of a man named Wang Zhi, who had disappeared in the mountains hundreds of years ago. (Jiahui 2017)

Scholars have read many morals into the story, including the unforgiving passage of time or even the danger of Go as a distraction;

however, according to the philosophy of Go, no second spent with the board can count as distraction. What the story of Wang Zhi invokes is the significance of dedication, or a desire to improve through infinite distractedness, if you will: it points toward the virtues of being lost in the game rather than to a maladaptive withdrawal from the environment.

Interestingly, Google's experiments with productive distraction come at a time when psychologists and neuroscientists are increasingly focusing on depathologizing nonconscious or "unconstrained" thought. Researchers estimate that we spend more than half of our waking life engaged in mind wandering or day dreaming, which have long figured as entirely self-related, hopelessly distracted, and potentially pathological mental states but have gradually been rehabilitated through resting state research in cognitive neuroscience. Depending on context, mind wandering has been categorized as task-unrelated, spontaneous, unintentional, meandering, intrinsic, stimulus-independent, and therefore distracted thought; at the same time, it is connected to goal-directed thinking, future planning, creativity, and an augmented social life. Mind wandering now figures "as a heterogeneous, fuzzy-boundaried construct that coheres amid patterns of overlapping and nonoverlapping features" (Seli et al. 2018, 479).

Whereas maladaptive mind wandering and failures in the retrieval of episodic memory have been linked to schizophrenic spectrum disorder, mood disorders, attention-deficit disorder, and last but not least, autism spectrum disorder, day dreaming in a controlled fashion is becoming an imperative in pursuit of an orderly social life and a successful career in today's strongholds of creative labour (Poerio and Smallwood 2016).[3] To put it bluntly, mind wandering can purportedly be disruptive and take the form of depressive rumination *or* support imaginative thinking and bring out the best in us (Alderson-Day and Callard 2016; Smallwood and Andrews-Hanna 2013).

Where productivity and pathology figure as two poles in the executions of a single program, the avoidance of disorder turns into a problem of regulation, adaptation, and optimization. Google's neuroscience-inspired AI is considered to provide an abundance of models that act as examples of what could go wrong; at the same time, they promise to illuminate "what the brain *should* compute, in order to provide the basis for successful behavior" (Kietzmann, McClure, and Kriegeskorte 2019, 9, emphasis added). Thus problems that typically concern only the neurosciences and psychology suddenly appear to have become a domain of engineering and computing.

The increasing prominence of unconstrained cognition and mind wandering in cognitive neuroscience is to a certain degree an effect of the methodological flipping of contrast that resting state research in brain imaging has brought about. Yet it has caught the interest of psychologists, in particular, thanks to its apparent significance for pathologies that arise preferably within informatically dense, contemporary urban environments. Avoiding maladaptive forms of mind wandering has thus turned into a problem of regulation and control through mental training. David Vago and Fadel Zeidan (2016) state, "A sense of peace and quiet in the mind is proposed to arise through mental training in concentration, nonconceptuality, and discernment, in contrast to the untrained, frenetic restlessness of mental time travel that is characteristic of daily activity in the postmodern setting. The frenetic resting state and associated brain network dynamics are believed to help scaffold attention and emotion throughout everyday waking life, but with the potential to interfere with cognitive performance, mood and affect when mind wandering occurs in the context of cognitive demand" (108).

Google DeepMind's interest in "episodic control" (Pritzel et al. 2017; Ritter et al. 2018) sticks out in this regard, for it is promoted as a technique that allows restless information processing algorithms to be consolidated through memory replay, which resembles episodes of controlled mind wandering. "Subconscious" processing of vivid memories is supposed to protect the algorithm from "catastrophic forgetting" that might arise due to an inevitable information overload (Kirkpatrick et al. 2017).

In this process, a novel laboratory has appeared on the scene that suggests the differentiation between productive distraction or frenetic restlessness is about to be conceived in the absence of thinking and acting human subjects on the level of algorithms that the brain might or might not successfully execute. In December 2018, Google open-sourced its flagship program, a fully 3-D game-like platform tailored for agent-based AI research. DeepMind Lab is observed from a first-person viewpoint and through the eyes of a simulated agent who encounters scenes rendered in rich, science-fiction-style visuals (Beattie et al. 2016b). However, since it did not seem likely that the average person would have achieved the same level of general intelligence unless we had "grown up in a world that looked like Space Invaders or Pac-Man," Google DeepMind opened another lab to the algorithmic public (Beattie et al. 2016a).

PsychLab takes the setup typically used in human psychology experiments and recreates it inside the virtual DeepMind Lab environment, thus allowing humans and artificial agents to take the same tests in order to minimize experimental differences. Google DeepMind researchers maintain that they hope "to facilitate a range of future such studies that simultaneously advance deep reinforcement learning and improve its links with cognitive science" (Leibo et al. 2018, 1).

Neuroscience-inspired AI is experimental in its capacity to bring something into existence, namely a "hazy geography of ... intelligence" (Mattern 2016) and a liminal space, where rethinking cognition figures as an exercise in speculation. Whether in the brain or in the Cloud, "statistics and geometry are linked in seeking to unearth recurring patterns that are scalable in time and space and can allow speculation on the future forms of population, their actions, and the shape of the space they will occupy" (Halpern et al. 2013, 296). Contemporary neural media hover above or in between the brain and the Cloud, providing models of how we could flourish or survive in the automated landscapes that smart city models and machine learning territories promote.

Is neuroscience-inspired AI thus about to take over the tasks of psychology and the social sciences? This is perhaps the wrong question to ask since it diverts attention to the imaginary potential of what, as of yet, is a rather mundane experimental protocol. Nevertheless, there appear to be striking parallels between an operation that aims to "look into" the minds of deep neural networks and mid-twentieth-century attempts to design "introspectometers" for social science research. Rebecca Lemov (2010) argues that Robert Merton's methodological thought (see Merton, Fiske, and Kendall 1956) translates well into the fantasies that surrounded early functional brain imaging. Just as Merton conceived the focus group as a preliminary stage in the design of a device to access a person's experiential stream of reality, fMRI was for quite some time expected to become such a mind-penetrating machine. The existential devices of neuroscience-inspired AI now iterate such fantasies in the absence of behaving humans through algorithmic entities that try to gradually live up to the cognitive capacities of their human others without mimicking their brains.

PSYCHIC LIFE IN THE CLOUDS

I would therefore like to add another dimension to what Des Fitzgerald (2018) describes as a "limit sociology" (18–20), which arranges "social life as an assemblage that now is as amenable to app-mediated psychological self-ratings as it is to ethnographic field-notes, and epidemiological surveys, and perhaps even, in future work, to direct biological measures" (21). Would AI that is inspired by neuroscience as well as cognitive neuroscience that is powered by AI, with their "flights of fancy" in regard to a sort of algorithmic sociality and their comparatively very mundane experimental protocols, not have to be part of this endeavour? One does not need to "believe" in emerging imaginaries in order to understand that, within this still developing domain, what we have come to know as being intelligent, social, and creative is increasingly up for grabs.

Whereas clouds, as John Durham Peters observes, "exist by disappearing" (interviewed in Hanrahan 2015), the cognitive algorithms that roam their hazy confines are here to stay. As they will continue to unsettle traditional ontologies of human and machine, we are called upon to find ways "to initiate thinking from within the machine and from within the very logic of the instrument" without subscribing to the imperatives "of capture, classification, and control" (Majaca and Parisi 2016). However, there is no reason to panic. As Fitzgerald (2018) observes, we should conceive "the material transformations of the present not as a cause for alarm, nor the sign of crisis, nor the absence of a future, nor still a premonition of death. But rather as a nudge to think less conventionally about just these kinds of binaries; even to let go [of] the sense of doom, to dial down the panic, to breathe in the forest air, to be calm" (21–2). Ever more catastrophic scenarios about how the life sciences or engineers are about to marginalize humanistic ideals might blind us to the opportunities that (patho)logical cognitive algorithms potentially hold.

What we see in this process is the human in motion, and observing its metamorphoses will help us to reconceive what we want it to be. Observing Google's neuroscience-inspired AI on one end of the emerging spectrum might put us in the position to occupy the liminal space that emerges at another end. It might indeed benefit our social lives if we leave behind understandings of human intelligence that represent no more than "a function of its situated emergence within a specific biochemical planetary condition from which it cannot escape without becoming something else" (Bratton 2019, 21). It might contribute to leaving behind the grids of power-knowledge that the human sciences of the global north have helped to build, if the question about what the human is occurs daily now in the work of engineers, apart from the dualism of biology and the social that is still being reproduced today in many discussions about "the human" (Rees 2018).

Of course, it is anything but given, or even likely, that this liminal space will indeed be occupied in any way similar to what I have briefly sketched out. Issues that have arisen over racial and gender bias inherent in algorithmic systems throughout the past years do not inspire confidence in the idea of putting social, and thus also political, issues into the hands of engineers and programmers (Broussard 2018; Bucher 2018; Noble 2018). Yet these issues have also persisted through periods when our beloved humanisms seemed to sit at the core of political initiatives. I believe that if the brain can be *studied with* engineers and programmers and through "experimental entanglement" (Fitzgerald and Callard 2015), we may be able to avoid merely recounting the deviations of post-anthropocentric intelligence from humanist perspectives. The uncertainties introduced through unbearably technical accounts of the psyche might productively unsettle humanisms that have never worked in our favour.

A critical "anthropology of the machine" (Fisch 2018) will have liberating effects if it is open to abandoning the emphatic attempts to differentiate between human and machine as well as that between human intelligence and artificial intelligence; instead, it must attend to how the distinction is mobilized in ever different ways and to different effect. Doing so requires, for instance, tracing and thinking through the very psychological concepts and cognitive states that inspire contemporary artificial intelligence research not merely to understand the idea behind certain algorithmic systems but also to gauge the effects that the circuits of existential systems might have on how we conceive of mental disorder and deviation from psychological (or sociological) norms. We are tasked with making the clouds of data derived from specific experimental constellations and protocols in neuroscience and psychology disappear again and again in order to see what emerges through the mingling of psychological concepts, statistical techniques, computational epistemologies, and coding practices – for if psychic life is to be colonized by the rhythms, waves, and patterns of machines, we should make sure that it is, in this very process, infinitely queered and diversified.[4]

Notes

1 Although such caveats appear legitimate at a time of admittedly abundant, mass-media-fuelled neurohype (Cooter 2014; see also Cromby, Newton, and Williams 2011), they often seem to overstate the neurosciences' actual potential reach. Somewhat ironically, neuroethics, a discipline entrusted with drawing lines in the sand of neuroscientific omniscience, has manifested the significance of the brain in the construction of the social subject (Giordano and Gordijn 2010; Illes and Bird 2006; Racine and Costa-von Aesch 2011; Racine and Zimmerman 2012). Although supplementing a discipline with a dedicated ethics branch will ever and involuntarily strengthen its claims to a certain degree, analysts and critics of the neurosciences have found neuroethicists and sociologists in particular guilty of overstating the disruptive potential of the neurosciences due to a marked preference for propositional rhetorics and promissory accounts of future opportunities (Brosnan 2011; Conrad and Vries 2011; De Vries 2007; Pitts-Taylor 2014).

2 The link between (social) cognition and social behaviour was first explicitly mentioned by Leslie Brothers in her seminal paper "The Social Brain" (1990), published in *Concepts of Neuroscience*. Brothers combines the notion that the brain commands automatic mechanisms designed for mastering social situations with the idea that these mechanisms remain open to modulation through social interaction. Her understanding of sociality as resulting from "the accurate perception of the dispositions and intentions of other individuals" (28) has been empirically enforced by the discovery of the so-called mirror system in the brain.

Mirror neurons were first reported in the macaque monkey's ventral premotor area by Giuseppe Di Pellegrino and colleagues (1992), although named as such only in a subsequent paper by the same group based at the University of Parma (Gallese et al. 1996). They reported that neurons in this particular area fired not only when the monkey performed a certain action but also when it observed another monkey or human experimenter perform the same action. The authors of the study speculated that the aptly named mirror neurons play an important role in understanding and anticipating the behaviour of others in social situations – a conclusion supported by the experimental observation that the same groups of neurons also fired when the experimenters conducted only mimicking movements.

The initially cautious hypotheses about social learning and its neurobiological correlates soon gave way to more confident speculations, which peaked in the first decade of the twenty-first century. Vittorio Gallese (2003), a leading member of the Parma group, stated that assuming his theory is correct, "a single mechanism – embodied simulation – can provide a common functional framework for all apparently different aspects of interpersonal relations" (521), including so-called mentalizing, or the simulation of emotions, feelings, and sensations. Understanding what others feel, or empathy in socio-psychological terms, became a "largely automatic process by which we 'read' the mental states of others" (Frith 2007, 671) through the activation of key brain structures.

Due to a lack of empirical proof for the existence of these brain structures, however, the ranks of those who doubt the significance or even existence of a mirror neuron system in the human brain have been swelling (Borg 2007; Dinstein et al. 2008; Lingnau, Gesierich, and Caramazza 2009) and the concept has gradually lost appeal, at least within the neurosciences (Kilner and Lemon 2013). Although mirror neurons have quickly changed from the most promising avenue in research on human empathy into "the most hyped concept in neuroscience" (Jarrett 2012), the socio-psychological or cognitive framing of social brains remains remarkably persistent.

3 Contributing to an increase in these concerns has been the prominence of the brain in many distinct discourses that involve economics, marketing, and the law, among others (Ariely and Berns 2010; Buckholz and Faigman 2014; Jones et al. 2013; Pickersgill and van Keulen 2011; Schneider and Woolgar 2015a, 2015b).

4 Meanwhile, even those who have advocated the sort of rigorous empiricism that brain imaging has come to signify are starting to doubt the

effectiveness of traditional neuroscience technologies and methodologies. In a 2017 interview with *Wired* magazine, Tom Insel, former director of the National Institute of Mental Health (NIMH) in the United States, admits that he "spent 13 years at NIMH really pushing on the neuroscience and genetics of mental disorders, and when I look back on that I realize that while I think I succeeded at getting lots of really cool papers published by really cool scientists at fairly large costs – I think $20 billion – I don't think we moved the needle in reducing suicide, reducing hospitalizations, improving recovery for the tens of millions of people who have mental illness" (quoted in Rodgers 2017).

5　The two studies mentioned were in fact follow-ups to an earlier study of Deena Weisberg and colleagues (2008), which was designed to test neuroscientific information in general with regard to its "seductive allure" and which caused a debate about the question of whether the publication of neuroscientific complements in popular science media could trick readers into uncritically adopting contested scientific knowledge. In their experiment, lay persons, neuroscience students, and neuroscience experts were provided with brief descriptions of psychological phenomena followed by one of four types of explanation, two of which contained irrelevant neuroscientific information. Since subjects from the two nonexpert groups considered the explanation supplemented by irrelevant results of brain imaging studies more satisfying, no matter whether the explanation was judged good or bad by experts, the authors of the study reasoned that it might indeed be neuroscientific jargon that is inappropriately persuasive.

However, the results of both studies did not remain uncontested. Five years and hundreds of uncritical citations later, Martha Farah and Cayce Hook (2013) of the Centre for Neuroscience and Society at the University of Pennsylvania claimed that the seductive allure of brain images and superfluous neuroscientific information was primarily an effect of bad experimental design. Citing three studies (Gruber and Dickerson 2012; Hook and Farah 2013; Michael et al. 2013) that fail to provide evidence for the seductive allure of brain images, they attribute the remarkable success of especially Weisberg and colleagues' study to their article's sensationalistic title, "The Seductive Allure of Neuroscience Explanations" (2008), as well as to the putative plausibility of the claim, which is considered to undermine a more critical examination of the study design. Farah and Hook (2013) suggest that the seductive allure of neuroscience explanations in Weisberg and colleagues' study might in fact be an effect of the considerably bulkier neuroscientific explanations. They further criticize David McCabe and Alan Castel (2008) for using diagrams and bar graphs that provide much less information about the topic than the much more detailed brain images: the surplus detail of the presented topographical maps might, so the authors argue, invalidate the comparison and thus the results of the study. In response to the criticism, Weisberg and her team conducted new experiments that ruled out factors that might limit the scope of their claims, and they came to the conclusion that the seductive allure is in fact a "reductive allure" (Hopkins, Weisberg, and Taylor 2016; Weisberg, Taylor, and Hopkins 2015; see also Neurocritic 2012).

6 Liv Hausken, Bettina Papenburg, and Sigrid Schmitz (2018) provide a valuable overview of the technicalities of disciplining volunteers' bodies in biomedical imaging.

7 Whereas many scholars from the realm of the social sciences typically encounter a reductionist laboratory situation that could potentially undermine more comprehensive notions of sociality, others see potential for an open-ended deliberation about what "the social" should – experimentally – come to involve. Felicity Callard and Des Fitzgerald (2015), for instance, advocate an experimental entanglement of the social sciences and neurosciences since more technical accounts of sociality might act as a corrective to the unforgivingly qualitative and therefore no less exclusionary understandings of sociality that the social sciences have cultivated over the course of the past century. Neuroscience commands experimental technologies that potentially etch together "local politics, de-oxygenated blood, sick bodies, nuclear physics, and the clinical gaze" (11) and that might therefore contribute to correcting the materialist blind spot of twentieth-century social sciences.

8 Neuroscientists Lisa Feldman Barrett and Ajay Satpute (2013) argue that what is at stake in the further development of the brain sciences is a "faculty psychology tradition" that is based on a modular understanding of the brain and has carved up human brain imaging research into at least three sister disciplines: affective, social, and cognitive neuroscience. They suggest that recent brain imaging studies of social brains challenge the modularity not only of the brain sciences but also and especially of the brain. By means of neuroscience techniques and technologies, the potentially obscuring experimental differentiation between cognition, affect, and sociality will be infused with productive ambivalence, inducing a multifaceted understanding of sociality and its putative obverse, namely the manifold concepts of anti-social behavior.

9 A prerequisite for such an engagement of neuroscientific methodology, however, is a more nuanced analysis of how a certain "reductionism" not only subtracts from but, time and again, also adds to the understandings

of the social that originate in its traditional habitats. "In other words, there is a big difference between an experiment that differentiates (social) groups purely on account of their different physiological attributes, and one that explores how those differences might be produced through social as well as physiological patternings and dynamics" (Callard and Fitzgerald 2015, 55). Scrutinizing a subject's experience in the scanner, for instance, still does not rank high on the neuroscientific agenda. Inspired by the methodologies of experimental psychology, which typically render subjects "anonymous and purportedly passive actors whose thoughts and 'behaviors' have been represented almost exclusively through experimenters' terms or numeric systems" (Morawski 2007, 129), the neurosciences tend to make use only of statistically normalized and anonymized data. Introspective reports from the side of the subject are often collected in debriefings, yet they are rarely used to validate the brain states that are turned into data by means of the scanner. In the service of reproducibility, the "phenomenology of a subject's experience is carefully excluded from the science of that subject's mind" (Callard and Fitzgerald 2014, 231).

10 The decisive push toward more tangible and measurable concepts of autism spectrum disorder, which resulted in what critics now call neurobiologization, did indeed come from the realm of the social sciences. Bonnie Evans (2018) points out that a diagnostic turn occurred in the 1960s, arguing that "autism grew up as resistance to a neoliberal agenda, a tool for sheltering certain people from the growing challenges of global capitalism" (see also Evans 2017). The drive to measure social "capacity" (Wing and Gould 1979) thus went hand in hand with a will to reform social scientific measurement and to promote social inclusion by clinical diagnoses of impairment, disorder, and disability.

Emphasizing the depersonalized and "machinic" aspects of the condition has been part of this process (Fein 2011) and has involved disentangling ASD from the psychoanalytical strongholds of language and meaning. Instead of focusing on life histories of adults diagnosed with autism, researchers have since prioritized the identification of biological markers in the brains of children who have not (yet) been diagnosed with ASD in order to find the autistic "real" within the brain. Autism has thus turned into a "disorder" of the brain and has been characterized as neurodevelopmental, which has entailed a shift in focus toward the prediagnostic and the infant "at risk" (Nadesan 2005, 64).

11 More information on the Visual 6502 project can be found at http://www.visual6502.org.

12 The Human Brain Project recently defined further research on episodic memory as one of its main strategic goals (see Amunts et al. 2016).

CHAPTER TWO

1 I am referring to the work of Geoffrey Bowker (2005) and his observation that data can never be "raw"; they merely turn into data *of* a phenomenon after they have been prepared to be digested. For an extended discussion of the "raw data" issue, see the section of this chapter titled "Maths."

2 A recent paper by Miles MacLeod and Nancy Nersessian (2018) provides insights into similar constellations and the effects on modelling practices in systems and computational biology.

3 See chapters 3 and 4 for a detailed account of the debates about the use of statistics in brain imaging.

4 See the work of Canay Özden-Schilling (2015, 2016) for a wonderful analysis of informational labour in electricity markets.

5 Natasha Myers (2015) observes a similar approach to imaging in the case of crystallographers.

6 The full passage from Kelly Joyce's (2008) book is worth reading: "As with NMR imaging, the practice of printing both numerical and visual forms of representations ended shortly after CT [computerized tomography] technology was placed under the jurisdiction of radiologists and radiology units. Calling attention to the 'radiology effect,' historian Bettyann Kevles (1997, 161) notes that 'Few, if any, radiologists ever looked at numbers,' and the practice of printing out numerical data ceased after the incorporation of CT scanners into radiology units. In the case of NMR imaging technology, the professional vision of radiologists caused a second change in data appearance by transforming the content of the image from multicolor to gray scale. The machine design was altered so that the final product was not a multicolor picture. Numerical values were coded into shades of gray, and the resulting images were printed on black and white film. NMR machines now produced the body as a black and white image, and representations of the body as arrays of reds, yellows, and blues disappeared from clinical practice" (39–40).

7 Nadine Levin (2014), for instance, argues that techniques for handling data are entangled with notions of complexity.

8 For further perspectives on outsiders' lives in the realm of neuroscience, see Paula Gould's (2004) article about physicists who help to decode the brain.

9 Rob Kitchin (2017) differentiates between two key translation challenges in the practice of coding: first, translating a task into a structured formula with an appropriate rule set (pseudo code) and, second, translating the pseudo code into source code that will perform the task or solve the problem after compiling.

10 On the backcover of her book *If ... Then: Algorithmic Power and Politics* (2018), Taina Bucher uses the term "powerful brokers of information" to describe contemporary pattern recognition algorithms, showing how they become part of and structure social practices. I am using it here, as detailed throughout the following chapters, to emphasize how various forms of data processing have been variously assigned to computers or lower-level information workers. See also the formidable history of early computation and intuitive management by Kira Lussier (2018).

CHAPTER THREE

1 The Loss-of-Confidence Project can be found at https://lossofconfidence. com. The initiators aim to have a peer-reviewed journal publish a paper that integrates various loss of confidence statements and that emphasizes the significance of publishing negative results that arise only during the afterlife of previous studies.

2 Publication biases were first analyzed by Harvard psychologist Robert Rosenthal (1979) under the moniker of "the file drawer problem," which represents an attempt to explain why negative results are often not published: they go straight into the file drawer. A group of researchers that includes Hal Pashler, one of Edward Vul's co-authors on the "Voodoo Correlations" article, has meanwhile set up a web archive called PsychFileDrawer at http://psychfiledrawer.org to solve the file drawer problem by encouraging and collecting replication attempts in experimental psychology. Leif Nelson, Joseph Simmons, and Uri Simonsohn (2018), however, argue that the file drawer problem does not accurately describe the replicability crisis: "Researchers would not so quickly give up on their chances for publication, nor would they abandon the beliefs that led them to run the study, just because the first analysis they ran was not statistically significant. They would instead explore the data further, examining, for example, whether outliers were interfering with the effect, whether the effect was significant within a subset of participants or trials, or whether it emerged when the dependent variable was coded differently" (514).

3 Opinions about the effect of p-hacking on scientific results are mixed. Meghan Head and co-authors (2015), for instance, believe that p-hacking is very common in scientific practice but argue that the effect seems to be weak relative to the actual effect sizes being measured. Chris Hartgerink (2017), however, found only ambiguous evidence for widespread p-hacking in his reanalysis of Head and colleagues' study. Such ambiguities and uncertainties as regards strategies for the management of uncertainties in data are characteristic of both the voodoo controversy and the replicability crisis in psychology, rendering their resolution even more complicated.

4 Based on an analysis of 3,081 cognitive neuroscience and psychology papers, Denes Szucs and John Ioannidis (2017) indeed write that "the recently reported low replication success in psychology is realistic, and worse performance may be expected for cognitive neuroscience" (1).

5 The blogs that reacted to the controversy – *Mind Hacks*, *The Neurocritic*, *Neuroskeptic*, and *Mutually Occluded*, to name but a few – are exclusively written by academics, albeit from entirely different disciplines: media studies, English literature, psychology, neuroscience, and cognitive science. In general, it can be said that the excitement was initially restricted to academics in the blogosphere but quickly spread to mainstream media.

6 See Svenja Matusall (2013) for a detailed analysis and critique of the concept of the social applied in social neuroscience with regard to experimental techniques and setup.

7 Neuroskeptic's twitter account (@neuro_skeptic) is a highly valuable source for discovering eroneous claims made in neuroscientific papers.

8 See note 2.

9 See note 1.

10 On his blog, Steve Luck (2018) of the UC Davis Center for Mind and Brain, for instance, admits that he has lost faith in p-values as a technique for managing experimental uncertainties. This novel uncertainty as regards an oft and commonly used statistical device was fuelled by a statement issued by the American Statistical Association in 2016 that contains six principles underlying the proper use and interpretation of the p-value and proposes to steer research into a post–"$p < 0.05$" era (Wasserstein and Lazar 2016, 131). A large group of leading researchers in various fields has thus proposed to raise the threshold of statistical significance by lowering the required p-value to 0.005 for publication (e.g., see Singh Chawla 2017). Leading statisticians such as Andrew Gelman (2016) of Columbia University in New York, however, argue that lowering the p-value will not have much of an effect if researchers do not change their attitude toward uncertainty and variation. The problem is that confidence

in the rejection of a competing hypothesis typically corresponds to an uncritical acceptance of the original hypothesis. Gelman explains, "Ultimately the problem is not with p-values but with the null-hypothesis significance testing, that parody of falsificationism in which straw-man null hypothesis A is rejected and this is taken as evidence in favor of pre-ferred alternative B ... [I]t seems to me that statistics is often sold as a sort of alchemy that transmutes randomness into certainty, an 'uncertainty laundering' that begins with data and concludes with success as measured by statistical significance." See chapters 3 and 4 for a more nuanced dis-cussion of the use of statistical devices for the management of uncertainty.

11 Jacob Cohen was a well-known American statistician who worked and published widely on statistical power analysis and effect size in the behavioural sciences.

INTERLUDE

1 Natasha Schull and Caitlin Zaloom's (2011) article in *Social Studies of Science* represents an inroad into discussions about neuroeconomics.

2 Meanwhile, however, different groups of researchers have started to caution against a methodical inflation of results (e.g., see Eklund, Nichols, and Knutsson 2016; Mueller et al. 2017).

CHAPTER FOUR

1 For further details, see Felicity Callard and Daniel Margulies's (2011) comprehensive history of the methodological transformation of the resting state into the brain's default mode.

2 See Amit Prasad (2007) and Britta Schinzel (2006) for accounts of "making bodies" in medical imaging.

3 Just like PET, fMRI has always been subject to criticism, for it represents no more than a proxy of "firing" neurons. In the case of fMRI, a consider-able lag between neuronal activity and its registration by the scanner through increased cerebral blood flow has been at the centre of attention: the blood oxygen level–dependent (BOLD) response does indeed occur roughly a second after a neuronal population has been "activated."

4 The International Consortium of Human Brain Mapping was founded in 1993 with a grant from the National Institute of Mental Health in the United States to work out a probabilistic reference system for the human brain. See Laboratory of Neuro Imaging, University of Southern California, at http://www.loni.usc.edu.

5 Categories for the selection of research participants are typically rooted in behavioural psychology and consist mainly of exclusion criteria; "that is to say, subjects are normal if they are untraumatized, unmedicated, unaddicted, non-diabetic, non-pregnant, and not having had any neurosurgery, psychiatric or neurological disorder" (Beaulieu 2001, 646). In an article titled "The Myth of the Normal, Average Human Brain" (Mazziotta et al. 2009), members of the International Consortium of Human Brain Mapping outline criteria and procedures for the selection of volunteers representative of the normal, average human across the adult lifespan. They propose three stages of selection: (1) telephone screening, (2) physical examination and in-person interviews, and (3) imaging.

  Criteria for excluding subjects from the study have been defined for every stage of the selection process. It is recommended, further, that subjects who pass both the telephone interviews and the physical examination should subsequently be scanned to identify possibly abnormal structures. In John Mazziotta and colleagues' (2009) study, only 180 out of 1,685 participants passed the three-stage examination and were admitted to the experimental stage. The *normal, average* brain is indeed the brain of a few.

6 The exclusion of cultural differences in cognitive neuroscience is an important issue that has led to the formation of cultural neuroscience, which promises to integrate the variations that occur in brains from different parts of the world. A recent paper in *Frontiers in Human Neuroscience* makes use of both the Montreal Neurological Institute's standard template and the statistical Chinese brain template, known as Chinese2020, and determines that the use of Chinese2020 for Chinese fMRI studies would improve the results (Shi et al. 2017). Whether cultural neuroscience will be able to solve the homogenization problem, however, remains to be seen. Critics fear an uncanny return of the race concept (e.g., see Malinowska 2016; Martinez Mateo et al. 2013) and/or, again, an inbuilt ignorance vis-à-vis the differences among cultural concepts in experimental paradigms (Roepstorff 2013).

7 See Johannes Stelzer and colleagues (2014) for a comprehensive review and critique of spatial normalization techniques in brain imaging.

8 The book MATLAB® *for Neuroscientists: An Introduction to Scientific Computing in* MATLAB® (Wallisch et al. 2014) indeed features a chapter titled "Neural Networks as Forest Fires: Stochastic Neurodynamics."

9 For a comprehensive overview of contemporary network neuroscience, see Danielle Bassett and Olaf Sporns (2017).

10 A similar hypothesis concerns the existence of "dark matter" (e.g., see Clancy 2015) or "dark neurons" (Shoham, O'Connor, and Segev 2006, 151)

within the brain. It differs from the dark energy hypothesis in that it concerns neurons that fire only occasionally and remain dormant for most of the time. The dark energy hypothesis, by contrast, targets neurons that have not been considered to be active in traditional brain imaging paradigms.

11 See David Merritt's article "Cosmology and Convention" (2017) for an analysis of the "dark energy" concept in physics. Only recently, a group of mathematicians of the University of Michigan and the University of California at Davis proposed that cosmic acceleration could be explained without reference to a fudge factor such as "dark energy" and within the theory of general relativity (Smoller, Temple, and Vogler 2017).

CHAPTER FIVE

1 This movement from tool to analogy is described by Tarja Knuuttila and Andrea Loettgers (2016) through the concept of artefactual models. Artefactual models differ from "fictional models" (e.g., see Frigg and Nguyen 2016) in that their fictionality does not derive from an "artificial reality" but is suggested through the use of particular general purpose devices that first allow researchers to bridge distinct systems.

2 *The Cambridge Handbook of the Neuroscience of Creativity* (Jung and Vartanian 2018), for instance, is full of references to the "default network" and to the "regional hubs" that purportedly govern its involvement in higher cognitive functions.

3 In the following, I concentrate on a methodological critique leveraged against the Human Brain Project. For a detailed account of how the review panel and the European Commission decided on the continuation of the project, see the highly interesting summary article by science writer and former molecular biologist Leonid Schneider (2017). Schneider maintains that the decision to continue the funding of the HBP was not taken based on scientific reports but subject to the bureaucratic changes that had been put in place.

4 The four selected projects are based at the Universities of Amsterdam, Naples, Glasgow, and Oslo and will investigate sleep and wakefulness, episodic memory, object representation, and the modulation of arousal levels (see https://www.humanbrainproject.eu/en/about/project-structure/subprojects).

5 Leonid Schneider (2018), a science journalist and outspoken critic of the Human Brain Project, has conducted an interview with the HBP's scientific director, Katrin Amunts, which addresses many of the criticisms that were leveraged against the consortium's approach.

6 The current interest in tracing "hidden" media infrastructure (Amoore
   2018; Parks and Starosielski 2015) speaks to the general absence of specific
   infrastructural elements in our cultural consciousness, particularly as
   regards digital communication. Only recently have humanities scholars
   begun to provide infrastructure with a new visibility in order to challenge
   their slide into an invisible background – a demise that is often politically
   volitional – thus broaching the issue that contemporary "media" are increasingly invisible *and* "blind" (e.g., see Allen and Bruder 2017; Gabrys 2016).

7 Tung-Hui Hu (2015) prominently references Steven Connor's book *The
   Matter of Air* (2010), where Connor writes about the decline of a belief in
   the ethereal as an inexhaustible reservoir for waste and exhaust: "The
   problem, not to say the disaster, is not the male 'finitizing' of the air, of
   which [Luce] Irigaray accuses [Martin] Heidegger and others, but the
   infinitizing of the air, the belief in the air as a horn of plenty, a bottomless
   fund of vastness, emptiness, openness, which can never be overdrawn. The
   air is not so much a way of having your cake and eating it, as of making
   your waste and never having to see it again" (275).

8 Originally a software concept, GPU-based computing has been optimized
   for pure number crunching in the context of high throughput tasks that
   execute a few very simple algorithms on very large amounts of data. While
   a classic central processing unit is optimized for serial processing and is
   highly versatile as regards the algorithms that it can run, its performance
   in data analysis is limited by slow memory access and by its small number
   of cores. GPUs, by contrast, typically have many cores that work in parallel and have very fast memory access, yet they can execute only very simple tasks. The currently very popular deep learning algorithms, for
   instance, work much more effectively on GPU-based systems, which means
   that a lot of resources are channelled into their development and availability via Cloud infrastructure.

   Examples of neurocomputation include the Neurogrid chip TrueNorth,
   developed by Kawbena Boahen and his colleagues at Stanford University's
   IBM lab in cooperation with the US Defense Advanced Research Projects
   Agency. Although the Neurogrid system combines analog computation to
   emulate ion channel activity and digital communication to simulate functional connectivity, TrueNorth is an actual neuromorphic chip with 4.096
   cores that support about 1 million brain cells and 256 million synapses
   (Merolla et al. 2014). Although said computing solutions are technically
   not simulations, Thomas Lippert, the director of the Institute for
   Advanced Simulation in Germany, notes that they might impact the
   "simulability" of certain models in the future (quoted in Pias 2011, 32).

CHAPTER SIX

1 Stefan Helmreich (2000) employs the term "liminal" (72) to describe the uncertain "nature" (80) of computer simulations.

2 Various open access tools meanwhile even allow researchers to convert MATLAB code directly into Python or provide a MATLAB-like interface within the Python ecosystem. For instance, NumPy (http://www.numpy.org), SciPy (https://www.scipy.org), and Matplotlib (https://matplotlib.org) provide Python-based ecosystems for mathematics, science, and engineering respectively.

3 Python derives further advantages from its openness. MATLAB, by contrast, is a proprietary and costly software that obscures the code of many algorithms and puts restrictions on code portability; to run an application coded in MATLAB, a computer needs to be licensed or at least to have the exact same version of the MATLAB Component Runtime installed.

4 Many universities have in recent years started to integrate respective courses into their curricula (Dreyfuss 2017); further, the Mozilla Foundation, for instance, has implemented boot camps for scientists willing to learn coding (Van Noorden 2014), and Python developers have started to compile core packages for different disciplines in order to facilitate what Jessica McKellar, a co-director of the Python Software Foundation, calls a "virtuous cycle," where new users extend the reach of their favourite programming language into new areas (quoted in Perkel 2015).

5 In his contribution to Benjamin Peters edited collection *Digital Keywords: A Vocabulary of Information Society and Culture*, Tarleton Gillespie (2016) describes chasing the etymology of the word "algorithm" as "chasing a ghost" and thus considers algorithms to be a "ghostly placeholder upon which computational systems now stand" (18–19). Fittingly, in a paper titled "Ghosts in Machine Learning for Cognitive Neuroscience: Moving from Data to Theory" (2017), neuroscientists Thomas Carlson and colleagues explain that because the challenges of applying machine learning methods to brain imaging data "metaphorically 'haunt'" their efforts, they refer to these challenges "as 'ghosts'" (88).

6 Friston's understanding that Bayesian statistics and deep leaning are inextricably entangled is not shared by all experts. Arguing that Bayesian methods constitute a mere belief system, Carlos Perez (2017) quotes Max Welling, a research chair in machine learning at the University of Amsterdam, who observes that "there are cultural differences between the two fields: where statistics is more focussed on statistical inference, that is, explaining and testing properties of a population from which we see a

random sample, machine learning is more concerned with making predictions, even if the prediction can not be explained very well (a.k.a. 'a black-box prediction')."

DEBRIEFING

1 Nick Seaver (2017), for instance, describes the algorithms that "the public and most critics focus on" as "distributed, probabilistic, secret, continuously upgraded, and corporately produced" (3), which turns them into opaque and elusive entities, as Jenna Burrell (2016) observes. For instance, since these technologies cannot be separated from their use in actuarial techniques (Harcourt 2007), "social values are encoded into mathematical processes and automated through techniques that scale normative logic" (Elish and Boyd 2018, 58). Frank Pasquale (2015) accordingly speaks of "the black box society," emphasizing the inscrutable algorithmic decisions taken in Silicon Valley or by Wall Street firms; Miriam Posner (2018), in an essay for *Logic* magazine, observes the opaque algorithmic operations of logistics; and both Virginia Eubanks (2018) and Safiya Umoja Noble (2018) tackle inequalities and racist practices that are being reinforced by and through high-tech profiling tools and search engines.

2 A more detailed and highly instructive account of the difference between "states" and "traits" in the history of psychology can be found in Nikolas Rose and Joelle Abi-Rached's book *Neuro: The New Brain Sciences and the Management of the Mind* (2013).

3 A recent paper, published in the journal *Cerebral Cortex*, links autism spectrum disorder to a malfunction in episodic memory retrieval due to reduced functional connectivity of the hippocampus with other large-scale neural networks (Cooper et al. 2017).

4 Hillary Powell, Hazel Morrison, and Felicity Callard (2018) provide a wonderful example of what such an investigation – one that targets the productive diversification and queering of seemingly well-defined mental and cognitive states – might look like. Their text, which takes account of the making of an exhibition that assembled various understandings of mind wandering, also details the authors' own approaches to creating environments that elicit, facilitate, or support "flights of fancy."

References

Abbott, Alison. 2009. "Brain Imaging Studies under Fire." *Nature* 457: 245. https://www.nature.com/news/2009/090113/full/457245a.html.

Abeles, Moshe, Ad Aertsen, Silvia Arber, Philippe Ascher, Francesco Battaglia, Daphne Bavelier, Heinz Beck, et al. 2014. "Open Message to the European Commission Concerning the Human Brain Project." https://www.neurofuture.eu.

Abend, Gabriel. 2018. "The Love of Neuroscience: A Sociological Account." *Sociological Theory* 36, no. 1: 88–116.

Abi-Rached, Joelle M. 2008. "The New Brain Sciences: Field or Fields?" *Brain, Self and Society: Working Papers*, no. 2: 1–7. http://eprints.lse.ac.uk/27941.

Abraham, Tara H. 2003. "From Theory to Data: Representing Neurons in the 1940s." *Biology and Philosophy* 18, no. 3: 415–26.

Adrian, Edgar D. 1931. "Potential Changes in the Isolated Nervous System of Dytiscus Marginalis." *Journal of Physiology* 72, no. 1: 132–51.

– 1932. "The Activity of Nerve Fibres." Reprinted in *Nobel Lectures: Physiology or Medicine, 1922–1941*, ed. Nobel Foundation, 293–300. Amsterdam: Elsevier, 1965.

Adrian, Edgar D., and Frederik J. Buytendijk. 1931. "Potential Changes in the Isolated Brain Stem of the Goldfish." *Journal of Physiology* 71, no. 2: 121–35.

Adrian, Edgar D., and Rachel Matthews. 1928. "The Action of Light on the Eye, Part III: The Interaction of Retinal Neurones." *Journal of Physiology* 65, no. 3: 273–98.

Alcalá-López, Daniel, Jonathan Smallwood, Elizabeth A. Jefferies, Frank van Overwalle, Kai Vogeley, Rogier B. Mars, Bruce Turetsky, et al. 2017.

"Computing the Social Brain Connectome across Systems and States." *Cerebral Cortex* 28, no. 7: 2207–32. https://doi.org/10.1093/cercor/bhx121.

Alderson-Day, Ben, and Felicity Callard. 2016. "Altered States: Resting State and Default Mode as Psychopathology." In *The Restless Compendium: Interdisciplinary Investigations of Rest and Its Opposites*, ed. Felicity Callard, Kimberly Staines, and James Wilkes, 11–17. Basingstoke, UK: Palgrave Macmillan.

Allen, Jamie, and Johannes Bruder. 2017. "Earth Blind: The Dark Empiricism of Earth Sensing." Visual vignette exhibited at the Sensor Publics Workshop in Munich, 5–7 April, 1–7. https://sensorpublics.files.wordpress.com/2017/09/sensor-publics-on-the-politics-of-sensing-and-data-infrastructures1-2.pdf.

Amoore, Louise. 2018. "Cloud Geographies: Computing, Data, Sovereignty." *Progress in Human Geography* 42, no. 1: 4–24.

Amunts, Katrin, Christoph Ebell, Jeff Muller, Martin Telefont, Alois Knoll, and Thomas Lippert. 2016. "The Human Brain Project: Creating a European Research Infrastructure to Decode the Human Brain." *Neuron* 92, no. 3: 574–81. https://doi.org/10.1016/j.neuron.2016.10.046.

Anderson, Christopher J., Štěpán Bahník, Michael Barnett-Cowan, Frank A. Bosco, Jesse Chandler, Christopher R. Chartier, Felix Cheung, et al. 2016. "Response to Comment on 'Estimating the Reproducibility of Psychological Science.'" *Science* 351, no. 6277: 1037. https://doi.org/10.1126/science.aad9163.

Andreasen, Nancy C., Daniel S. O'Leary, Ted Cizadlo, Stephan Arndt, Karim Rezai, G. Leonard Watkins, Laura L. Boles Ponto, and Richard D. Hichwa. 1995. "Remembering the Past: Two Facets of Episodic Memory Explored with Positron Emission Tomography." *American Journal of Psychiatry* 152, no. 11: 1576–85.

Ariely, D., and G.S. Berns. "Neuromarketing: The Hope and Hype of Neuroimaging in Business." *Nature Reviews Neuroscience* 11, no. 4 (2010): 284–92.

Baker, Monya. 2016. "1,500 Scientists Lift the Lid on Reproducibility." *Nature* 533: 452–4. https://www.nature.com/news/1-500-scientists-lift-the-lid-on-reproducibility-1.19970.

Barad, Karen. 2007. *Meeting the Universe Halfway: Quantum Physics and the Entanglement of Matter and Meaning*. Durham, NC: Duke University Press.

Barrett, Lisa Feldman. 2009. "Understanding the Mind by Measuring the

Brain: Lessons from Measuring Behavior (Commentary on Vul et al., 2009)." *Perspectives on Psychological Science* 4, no. 3: 314–18.

Barrett, Lisa Feldman, and Ajay B. Satpute. 2013. "Large-Scale Brain Networks in Affective and Social Neuroscience: Towards an Integrative Functional Architecture of the Brain." *Current Opinion in Neurobiology* 23, no. 3: 361–72.

Bassett, Danielle, and Olaf Sporns. 2017. "Network Neuroscience." *Nature Neuroscience* 20, no. 3: 353–64.

Bates, David. 2014. "Unity, Plasticity, Catastrophe: Order and Pathology in the Cybernetic Era." In *Catastrophes: A History and Theory of an Operative Concept*, ed. Andreas Killen and Nitzan Lebovic, 32–54. Berlin: De Gruyter.

Beattie, Charles, Joel Z. Leibo, Stig Petersen, and Shane Legg. 2016a. "Open-Sourcing DeepMind Lab." *DeepMind*, 3 December. https://deepmind.com/blog/open-sourcing-deepmind-lab.

Beattie, Charles, Joel Z. Leibo, Denis Teplyashin, Tom Ward, Marcus Wainwright, Heinrich Küttler, Andrew Lefrancq, et al. 2016b. "DeepMind Lab." *arXiv*, 13 December, 1–11. https://arxiv.org/abs/1612.03801.

Beaty, Roger E., Mathias Benedek, Scott Barry Kaufman, and Pau J. Silva. 2015. "Default and Executive Network Coupling Supports Creative Idea Production." *Scientific Reports* 5, no. 10964: 1–14. https://www.nature.com/articles/srep10964.

Beaulieu, Anne. 2001. "Voxels in the Brain: Neuroscience, Informatics and Changing Notions of Objectivity." *Social Studies of Science* 31, no. 5: 635–80.

– 2002. "A Space for Measuring Mind and Brain: Interdisciplinary and Digital Tools in the Development of Brain Mapping and Functional Imaging, 1980–1990." *Brain and Cognition* 49: 13–33.

– 2003. "Brains, Maps and the New Territory of Psychology." *Theory and Psychology* 13, no. 4: 561–8.

Begley, Sharon. 2009. "Of Voodoo and the Brain." *Newsweek*, 30 January. https://www.newsweek.com/sharon-begley-voodoo-and-brain-77717.

Bell, Vaughan. 2008. "Voodoo Correlations in Social Brain Studies." *Mind Hacks*, 29 December. https://mindhacks.com/2008/12/29/voodoo-correlations-in-social-brain-studies.

Bennett, Craig M. 2009. "The Story behind the Atlantic Salmon." *Prefrontal.org: A Personal Weblog of Developmental Cognitive Neuroscience*, 18 December. http://prefrontal.org/blog/2009/09/the-story-behind-the-atlantic-salmon.

Bennett, Craig M., Abigail A. Baird, Michael B. Miller, and George L. Wolford. 2009. "Neural Correlates of Interspecies Perspective Taking in the Post-Mortem Atlantic Salmon: An Argument for Multiple Comparisons Correction." Poster at the 15th Annual Meeting of the Organization for Human Brain Mapping, San Francisco. *Prefrontal.org: A Personal Weblog of Developmental Cognitive Neuroscience*, 21 June. http://prefrontal.org/blog/2009/06/human-brain-mapping-2009-presentations.

Bennett, Craig M., Michael B. Miller, and George L. Wolford. 2010. "Neural Correlates of Interspecies Perspective Taking in the Post-Mortem Atlantic Salmon: An Argument for Multiple Comparisons Correction: Supplement 1." *Journal of Serendipitous and Unexpected Results* 1, no. 1: 1–5. https://teenspecies.github.io/pdfs/NeuralCorrelates.pdf.

Berger, Hans. 1929. "Über das Elektrenkephalogramm des Menschen." *Archiv für Psychiatrie und Nervenkrankheiten* 87, no. 1: 527–70.

– 1969. *Hans Berger on the Electroencephalogram of Man: The Fourteen Original Reports on the Human Electroencephalogram.* Trans. and ed. Pierre Gloor. Amsterdam: Elsevier.

Biswal, Bharat, F. Zerrin Yetkin, Victor M. Haughton, and James Hyde. 1995. "Functional Connectivity in the Motor Cortex of Resting Human Brain Using Echo-Planar MRI." *Magnetic Resonance in Medicine* 34, no. 4: 537–41.

Blackman, Lisa. 2019. "Haunted Data, Transmedial Storytelling, Affectivity: Attending to 'Controversies' as Matters of Ghostly Concern." *Ephemera: Theory and Politics in Organisation* 19, no. 1: 31–52.

Blume, Stuart S. 1992. *Insight and Industry: On the Dynamics of Technological Change in Medicine.* Cambridge, MA: MIT Press.

Borck, Cornelius. 2005. *Hirnströme: Eine Kulturgeschichte der Elektroenzephalographie.* Göttingen: Wallstein.

– 2008. "Recording the Brain at Work: The Visible, the Readable, and the Invisible in Electroencephalography." *Journal of the History of the Neurosciences* 17, no. 3: 367–79.

Borg, Emma. 2007. "If Mirror Neurons Are the Answer, What Was the Question?" *Journal of Consciousness Studies* 14, no. 8: 5–19.

Borrelli, Arianna. Forthcoming. "Program FAKE: Monte Carlo Event Generators as Tools of Theory in Early High Energy Physics." *NTM Zeitschrift für Geschichte der Wissenschaften, Technik und Medizin* 49.

Bowker, Geoffrey C. 1994. *Science on the Run: Information Management and Industrial Geophysics at Schlumberger, 1920–1940*. Cambridge, MA: MIT Press.

– 2005. *Memory Practices in the Sciences*. Cambridge, MA: MIT Press.

Bowker, Geoffrey, Karen Baker, Florence Millerand, and David Ribes. 2010. "Toward Information Infrastructure Studies: Ways of Knowing in a Networked Environment." In *International Handbook of Internet Research*, ed. Jeremy Hunsinger, Lisbeth Klastrup, and Matthew M. Allen, 97–117. Amsterdam: Springer.

Bowker, Geoffrey, and Susan Leigh Star. 1999. *Sorting Things Out: Classification and Its Consequences*. Cambridge, MA: MIT Press.

Brante, Thomas, Steve Fuller, and William Lynch. 1993. *Controversial Science: From Content to Contention*. Albany: State University of New York Press.

Bratton, Benjamin H. 2015. *The Stack: On Software and Sovereignty*. Cambridge, MA: MIT Press.

– 2019. "Future Trace Effects of the Post-Anthropocene." In *Machine Landscapes: Architectures of the Post-Anthropocene*, ed. Liam Young, 15–21. Hoboken, NJ: Wiley.

Brodmann, Korbinian. 1909. *Vergleichende Lokalisationslehre der Grosshirnrinde: In ihren Principien dargestellt auf Grundlage des Zellenbaus*. Leipzig: Johann Ambrosius Barth Verlag.

Brooks, Rodney. 1989. "A Robot That Walks: Emergent Behavior from a Carefully Evolved Network." *Neural Computation* 1, no. 2: 253–62.

– 1990. "Elephants Don't Play Chess." *Robotics and Autonomous Systems* 6, nos 1–2: 3–15. https://doi.org/10.1016/S0921-8890(05)80025-9.

– 2018. "Steps Toward Super Intelligence IV, Things to Work on Now." *Robots, AI, and Other Stuff*, 15 July. https://rodneybrooks.com/forai-steps-toward-super-intelligence-iv-things-to-work-on-now.

Brooks, Rodney, and Anita M. Flynn. 1989. "Fast, Cheap and Out of Control: A Robot Invasion of the Solar System." *Journal of the British Interplanetary Society* 42: 478–85.

Brooks, Rodney, Demis Hassabis, Denis Bray, and Amnon Shashua. 2012. "Turing Centenary: Is the Brain a Good Model for Machine Intelligence?" *Nature* 482: 462–3. https://www.nature.com/articles/482462a.

Brosnan, Caragh. 2011. "The Sociology of Neuroethics: Expectational Discourses and the Rise of a New Discipline." *Sociology Compass* 5, no. 4: 287–97.

Brothers, Leslie. 1990. "The Social Brain: A Project for Integrating Primate Behaviour and Neurophysiology in a New Domain." *Concepts in Neuroscience* 1: 27–51.

Broussard, Meredith. 2018. *Artificial Unintelligence: How Computers Misunderstand the World*. Cambridge, MA: MIT Press.

Bruder, Johannes. 2017. "Infrastructural Intelligence: Contemporary Entanglements between Neuroscience and AI." *Progress in Brain Research* 233: 101–29.

Bucher, Taina. 2016. "Neither Black nor Box: Ways of Knowing Algorithms." In *Innovative Methods in Media and Communication Research*, ed. Sebastian Kubitschko and Anne Kaun, 81–98. Cham, Switzerland: Palgrave Macmillan.

– 2018. *If ... Then: Algorithmic Power and Politics*. New York: Oxford University Press.

Buckholz, J.W., and D.L. Faigman. 2014. "Promises, Promises for Neuroscience and Law." *Current Biology* 24, no. 18: 861–7.

Buckner, Randy L., Jessica Andrews-Hanna, and Daniel L. Schacter. 2008. "The Brain's Default Network: Anatomy, Function, and Relevance to Disease." *Annals of the New York Academy of Sciences* 1124, no. 1: 1–38. https://doi.org/10.1196/annals.1440.011.

Burrell, Jenna. 2016. "How the Machine 'Thinks': Understanding Opacity in Machine Learning Algorithms." *Big Data and Society* 3, no. 1: 1–12. https://doi.org/10.1177%2F2053951715622512.

Burrington, Ingrid. 2016. *Networks of New York: An Illustrated Field Guide to Urban Internet Infrastructure*. Brooklyn, NY: Melville House.

Burton-Hill, Clemency. 2016. "The Superhero of Artificial Intelligence: Can This Genius Keep It in Check?" *Guardian* (London), 16 February. https://www.theguardian.com/technology/2016/feb/16/demis-hassabis-artificial-intelligence-deepmind-alphago.

Calhoun, Adam J. 2014. "The Members of the HBP Are Saddened by the Open Letter Posted on Neurofuture.eu (Updated x2)." *neuroecology*, 10 July. https://neuroecology.wordpress.com/2014/07/10/the-members-of-the-hbp-are-saddened-by-the-open-letter-posted-on-neurofuture-eu.

Callard, Felicity and Des Fitzgerald. 2014. "Experimental Control: What Does It Mean for a Participant to 'Feel Free'?" *Consciousness and Cognition* 27: 231–2. https://doi.org/10.1016/j.concog.2014.05.008.

– 2015. *Rethinking Interdisciplinarity across the Social Sciences and Neurosciences*. Basingstoke, UK: Palgrave Macmillan.

Callard, Felicity, and Daniel Margulies. 2010. "The Industrious Subject: Cognitive Neuroscience's Revaluation of 'Rest.'" In *Cognitive Architecture: From Biopolitics to Noopolitics – Architecture and Mind in the Age of Communication and Information*, ed. Deborah

Hauptmann and Warren Neidich, 324–46. Rotterdam: 010 Publishers.

– 2011. "The Subject at Rest: Novel Conceptualizations of Self and Brain from Cognitive Neuroscience's Study of the 'Resting State.'" *Subjectivity* 4, no. 3: 227–57.

Callard, Felicity, Jonathan Smallwood, and Daniel Margulies. 2012. "Default Positions: How Neuroscience's Historical Legacy Has Hampered Investigation of the Resting Mind." *Frontiers in Psychology* 3, art. 321: 1–6. https://www.frontiersin.org/articles/10.3389/ fpsyg.2012.00321/full.

Campbell, Nancy D. 2010. "Toward a Critical Neuroscience of 'Addiction.'" *BioSocieties* 5, no. 1, 89–104.

Canguilhem, Georges. 2008. *Knowledge of Life*. New York: Fordham.

Carlson, Thomas, Erin Goddard, David M. Kaplan, Colin Klein, and J. Brendan Ritchie. 2018. "Ghosts in Machine Learning for Cognitive Neuroscience: Moving from Data to Theory." *NeuroImage* 180, part A: 88–100. https://doi.org/10.1016/j.neuroimage.2017.08.019.

Carse, Ashley. 2017. "Keyword: Infrastructure: How a Humble French Engineering Term Shaped the Modern World." In *Infrastructures and Social Complexity: A Companion*, ed. Penny Harvey, Casper Bruun Jensen, and Atsuro Morita, 29–39. London: Routledge.

Chau, Wilkin, and Anthony R. McIntosh. 2005. "The Talairach Coordinate of a Point in the MNI Space: How to Interpret It." *NeuroImage* 25, no. 2: 408–16.

Choudhury, Suparna, Saskia Kathi Nagel, and Jan Slaby. 2009. "Critical Neuroscience: Linking Neuroscience and Society through Critical Practice." *BioSocieties* 4, no. 1: 61–77.

Clancy, Kelly. 2015. "Here's Why Your Brain Seems Mostly Dormant." *Nautilus*, no. 27, 6 August. http://nautil.us/issue/27/dark-matter/ heres-why-your-brain-seems-mostly-dormant.

– 2017. "A Computer to Rival the Brain." *The New Yorker*, 15 February. http://www.newyorker.com/tech/elements/a-computer-to-rival-the-brain.

Cohn, Simon. 2004. "Increasing Resolution, Intensifying Ambiguity: An Ethnographic Account of Seeing Life in Brain Scans." *Economy and Society* 33, no. 1: 52–76.

– 2008. "Making Objective Facts from Intimate Relations: The Case of Neuroscience and Its Entanglements with Volunteers." *History of the Human Sciences* 21, no. 4: 86–103.

Collins, Harry M., ed. 1981. "Knowledge and Controversy: Studies of Modern Natural Science." Special issue of *Social Studies of Science* 11, no. 1: 1–158.

Collins, Harry M., and T.J. Pinch. 1993. *The Golem: What Everyone Should Know about Science*. Cambridge, UK: Cambridge University Press.

Connor, Steven. 2010. *The Matter of Air: Science and Art of the Ethereal*. London: Reaktion Books.

Conrad, Eric C., and Raymond de Vries. 2011. "Field of Dreams: A Social History of Neuroethics." In *Advances in Medical Sociology: Sociological Reflections on the Neurosciences*, vol. 13, ed. Martyn Pickersgill and Ira van Keulen, 299–324. Bingley, UK: Emerald.

Cooper, Rose, Franziska R. Richter, Paul M. Bays, Kate C. Plaisted-Grant, Simon Baron-Cohen, and Jon S. Simons. 2017. "Reduced Hippocampal Functional Connectivity during Episodic Memory Retrieval in Autism." *Cerebral Cortex* 27, no. 2: 888–902.

Cooter, Roger. 2014. "Neural Veils and the Will to Historical Critique: Why Historians of Science Need to Take the Neuro-Turn Seriously." *Isis: A Journal of the History of Science* 105, no. 1: 145–54.

Cornel, Tabea. 2017. "Something Old, Something New, Something Pseudo, Something True: Pejorative and Deferential References to Phrenology since 1840." *Proceedings of the American Philosophical Society* 161, no. 4: 299–332.

Cromby, John, Tim Newton, and Simon J. Williams. 2011. "Neuroscience and Subjectivity." *Subjectivity* 4, no. 3: 215–26.

Dascal, Marcelo. 1998. "The Study of Controversies and the Theory and History of Science." *Science in Context* 11, no. 2: 147–54.

Daston, Lorraine. 2016. "Cloud Physiognomy." *Representations* 135, no. 1: 45–71.

Daunizeau, Jean, Kerstin Preuschoff, Karl Friston, and Klaas Stephan. 2011. "Optimizing Experimental Design for Comparing Models of Brain Function." *PLOS Computational Biology* 7, no. 11: 1–18. https://doi.org/10.1371/journal.pcbi.1002280.

Davanger, Syend. 2015. "The Brain's Default Mode Network – What Does It Mean to Us?" Interview with Marcus Raichle. *The Meditation Blog*, 9 March. https://www.themeditationblog.com/the-brains-default-mode-network-what-does-it-mean-to-us.

De Mol, Liesbeth. Forthcoming. "'A Pretence of What Is Not'? A Study of Simulation(s) from the ENIAC Perspective." *NTM Zeitschrift für Geschichte der Wissenschaften, Technik und Medizin* 49.

De Vos, Jan. 2016. *The Metamorphoses of the Brain: Neurologisation and Its Discontents*. London: Palgrave Macmillan.

De Vries, Raymond. 2007. "Who Will Guard the Guardians of

Neuroscience? Firing the Neuroethical Imagination." *EMBO Reports* 8: S65-S69.

Deco, Gustavo, Morten L. Kringelbach, Viktor K. Jirsa, and Petra Ritter. 2017. "The Dynamics of Resting Fluctuations in the Brain: Metastability and Its Dynamical Cortical Core." *Scientific Reports* 7, no. 3095: 1–14. https://doi.org/10.1038/s41598-017-03073-5.

Di Pellegrino, Giuseppe, Luciano Fadiga, Leonardo Fogassi, Vittorio Gallese, and Giacomo Rizzolatti. 1992. "Understanding Motor Events: A Neurophysiological Study." *Experimental Brain Research* 91, no. 1: 176–80.

Diener, Ed. 2009. "Editor's Introduction to Vul et al. (2009) and Comments." *Perspectives on Psychological Science* 4, no. 3: 272–3.

Ding, Xaio Pan, Si Jia Wu, Jiangang Liu, Kang Le, and Genyue Fu. 2017. "Functional Neural Networks of Honesty and Dishonesty in Children: Evidence from Graph Theory Analysis." *Scientific Reports* 7, no. 12085: 1–10. https://doi.org/10.1038/s41598-017-11754-4.

Dinstein, Ilan, Cibu Thomas, Marlene Behrmann, and David J. Heeger. 2008. "A Mirror up to Nature." *Current Biology* 18, no. 1: 13–18.

Doya, Kenji, Shin Ishii, Alexandre Pouget, and Rajesh P.N. Rao. 2006. *Bayesian Brain: Probabilistic Approaches to Neural Coding*. Cambridge, MA: MIT Press.

Dreyfuss, Emily. 2017. "Want to Make It as a Biologist? Better Learn to Code." *Wired*, 3 October. https://www.wired.com/2017/03/biologists-teaching-code-survive.

Dumit, Joseph. 2011. "Critically Producing Brain Images of Mind." In *Critical Neuroscience: A Handbook of the Social and Cultural Contexts of Neuroscience*, ed. Soupharna Choudhury and Jan Slaby, 195–225. Oxford: Wiley-Blackwell.

– 2016. "Plastic Diagrams: Circuits in the Brain and How They Got There." In *Plasticity and Pathology: On the Formation of the Neural Subject*, ed. David Bates and Nima Bassiri, 219–67. New York: Fordham University Press.

Dunkley, Benjamin T., Karolina Urban, Leodante Da Costa, Simeon M. Wong, Elizabeth W. Pang, and Margot J. Taylor. 2018. "Default Mode Network Oscillatory Coupling Is Increased Following Concussion." *Frontiers in Neurology* 9, art. 280: 1–11. https://doi.org/10.3389/fneur.2018.00280.

Easton, Alexander, and Nathan Emery, eds. 2012. *The Cognitive Neuroscience of Social Behaviour*. Hove, NY: Psychology Press.

Edwards, Paul. 2003. "Infrastructure and Modernity: Force, Time, and

Social Organization in the History of Sociotechnical Systems." In *Technology and Modernity: The Empirical Turn*, ed. Philip Brey, Arie Rip, and Andrew Feenberg, 185–225. Cambridge, MA: MIT Press.

– 2010. *A Vast Machine: Computer Models, Climate Data, and the Politics of Global Warming*. Cambridge, MA: MIT Press.

Ehrenberg, Alain. 2009. *The Weariness of the Self: Diagnosing the History of Depression in the Contemporary Age*. Montreal and Kingston: McGill-Queen's University Press.

– 2011. "The 'Social' Brain: An Epistemological Chimera and a Sociological Fact." In *Neurocultures: Glimpses into an Expanding Universe*, ed. Francisco Ortega and Fernando Vidal, 117–40. Frankfurt am Main: Peter Lang.

Eklund, Anders, Thomans E. Nichols, and Hans Knutsson. 2016. "Cluster Failure: Why fMRI Inferences for Spatial Extent Have Inflated False-Positive Rates." *Proceedings of the National Academy of Sciences* 113, no. 28: 7900–5.

Elish, M.C., and Danah Boyd. 2018. "Situating Methods in the Magic of Big Data and AI." *Communication Monographs* 85, no. 1: 57–80. https://doi.org/10.1080/03637751.2017.1375130.

Emery, Nathan. 2012. "The Evolution of Social Cognition." In *The Cognitive Neuroscience of Social Behaviour*, ed. Alexander Easton and Nathan Emery, 115–56. Hove, NY: Psychology Press.

Engelhardt, H.T., and Arthur L. Caplan. 1987. *Scientific Controversies: Case Studies in the Resolution and Closure of Disputes in Science and Technology*. Cambridge, UK: Cambridge University Press.

Ensmenger, Nathan. 2012. "Is Chess the Drosophila of Artificial Intelligence? A Social History of an Algorithm." *Social Studies of Science* 42, no. 1: 5–30. https://doi.org/10.1177/0306312711424596.

Eubanks, Virginia. 2018. *Automating Inequality: How High-Tech Tools Profile, Police, and Punish the Poor*. New York, NY: St Martin's.

European Commission. 2015. "The Human Brain Project Flagship: 1st Technical Project Review." http://ec.europa.eu/information_society/newsroom/cf/dae/document.cfm?action=display&doc_id=8923.

Evans, Bonnie. 2017. *The Metamorphosis of Autism: A History of Child Development in Britain*. Manchester, UK: Manchester University Press.

– 2018. "The Autism Paradox." *Aeon*, 8 January. https://aeon.co/essays/the-intriguing-history-of-the-autism-diagnosis.

Fanelli, Daniele. 2010. "'Positive' Results Increase down the Hierarchy of the Sciences." *PLOS ONE* 5, no. 4: 1–10. https://doi.org/10.1371/journal.pone.0010068.

– 2012. "Negative Results Are Disappearing from Most Disciplines and Countries." *Scientometrics* 90, no. 3: 891–904.

Farah, Martha J., and Cayce J. Hook. 2013. "The Seductive Allure of 'Seductive Allure.'" *Perspectives on Psychological Science* 8, no. 1: 88–90.

Fein, Elizabeth. 2011. "Innocent Machines: Asperger's Syndrome and the Neurostructural Self." In *Advances in Medical Sociology*, vol. 13, *Sociological Reflections on the Neurosciences*, ed. Martyn Pickersgill and Ira van Keulen, 27–49. Bingley, UK: Emerald.

Fernandez, G., K. Specht, S. Weis, I. Tendolkar, M. Reuber, J. Fell, P. Klaver, J. Ruhlmann, J. Reul, and C. E. Elger. 2003. "Intrasubject Reproducibility of Presurgical Language Lateralization and Mapping Using fMRI." *Neurology* 60, no. 6: 969–75.

Filevich, Elisa, Patricia Vanneste, Marcel Bass, Wim Fias, Patrick Haggard, and Simone Kühn. 2013. "Brain Correlates of Subjective Freedom of Choice." *Consciousness and Cognition* 22, no. 4: 1271–84.

Fisch, Michael. 2018. *An Anthropology of the Machine: Tokyo's Commuter Train Network*. Chicago: University of Chicago Press.

Fitzgerald, Des. 2017. *Tracing Autism: Uncertainty, Ambiguity and the Affective Labor of Neuroscience*. Seattle: University of Washington Press.

– 2018. "What Was Sociology?" *History of the Human Sciences*: 1–29. http://orca.cf.ac.uk/id/eprint/112960.

Fitzgerald, Des, and Felicity Callard. 2015. "Social Science and Neuroscience beyond Interdisciplinarity: Experimental Entanglements." *Theory, Culture and Society* 32, no. 1: 3–32.

Fleck, Ludwik. 1979. *Genesis and Development of a Scientific Fact*. Chicago: University of Chicago Press.

Fleming, Steve. 2016. "False Functional Interference: What Does It Mean to Understand the Brain?" *The Elusive Self*, 29 May. https://elusiveself.wordpress.com/2016/05/29/false-functional-inference-what-does-it-mean-to-understand-the-brain.

Fox, Kieran C., R. Nathan Spreng, Melissa Ellamil, Jessica R. Andrews-Hanna, and Kalina Christoff. 2015. "The Wandering Brain: Meta-Analysis of Functional Neuroimaging Studies of Mind-Wandering and Related Spontaneous Thought Processes." *NeuroImage* 111: 611–21.

Fox, Peter T., Joel S. Perlmutter, and Marcus E. Raichle. 1984. "Stereotactic Method for Determining Anatomical Localization in Physiological Brain Images." *Journal of Cerebral Blood Flow and Metabolism* 4, no. 4: 634.

Frackowiak, Richard. 2014. "Defending the Grand Vision of the Human Brain Project." *New Scientist*, no. 2978, 19 July. https://www.newscientist.com/article/mg22329784-400-defending-the-grand-vision-of-the-human-brain-project.

Frégnac, Yves, and Gilles Laurent. 2014. "Neuroscience: Where Is the Brain in the Human Brain Project?" *Nature* 513: 27–9. https://www.nature.com/news/neuroscience-where-is-the-brain-in-the-human-brain-project-1.15803.

Frigg, Roman, and James Nguyen. 2016. "The Fiction View of Models Reloaded." *The Monist* 99, no. 3: 225–42.

Friston, Karl. 2010a. "The Free-Energy Principle: A Unified Brain Theory?" *Nature Reviews Neuroscience* 11, no. 2: 127–38.

– 2010b. "Is the Free-Energy Principle Neurocentric?" *Nature Reviews Neuroscience* 11, no. 8: 605.

– 2012. "The History of the Future of the Bayesian Brain." *Neuroimage* 62, no. 2: 1230–3.

Frith, Chris D. 2007. "The Social Brain?" *Philosophical Transactions of the Royal Society B Biological Sciences* 362, no. 1480: 671–8.

Fyles, Matt. 2017. "Inside an AI 'Brain' – What Does Machine Learning Look Like?" *Graphcore*. https://www.graphcore.ai/posts/what-does-machine-learning-look-like.

Gabrys, Jennifer. 2016. *Program Earth: Environmental Sensing Technology and the Making of a Computational Planet*. Minneapolis: University of Minnesota Press.

Galison, Peter L. 1996. "Computer Simulations and the Trading Zone." In *The Disunity of Science: Boundaries, Contexts, and Power*, ed. Peter Galison and David J. Stump, 118–57. Stanford, CA: Stanford University Press.

Gallese, Vittorio. 2003. "The Manifold Nature of Interpersonal Relations: The Quest for a Common Mechanism." *Philosophical Transactions of the Royal Society B Biological Sciences* 358, no. 1431: 517–28.

Gallese, Vittorio, Luciano Fadiga, Leonardo Fogassi, and Giacomo Rizzolatti. 1996. "Action Recognition in the Premotor Cortex." *Brain: A Journal of Neurology* 119, no. 2: 593–609.

Gazzaniga, Michael S. 2005. *The Ethical Brain*. New York: Dana.

Gelly, Sylvain, Levente Kocsis, Marc Schoenauer, Michèle Sebag, David Silver, Csaba Szepesvari, and Olivier Teytaud. 2012. "The Grand Challenge of Computer Go: Monte Carlo Tree Search and Extensions." *Communications of the ACM* 55, no. 3: 106–13.

Gelman, Andrew. 2016. "The Problems with P-Values Are Not Just with

P-Values: My Comments on the Recent ASA Statement." *Statistical Modelling, Causal Inference, and Social Science*, 7 March. https://statmodeling.stat.columbia.edu/2016/03/07/29212.

Gerlach, Kathy D., R. Nathan Spreng, Kevin P. Madore, and Daniel L. Schacter. 2014. "Future Planning: Default Network Activity Couples with Frontoparietal Control Network and Reward-Processing Regions during Process and Outcome Simulations." *Social Cognitive and Affective Neuroscience* 9, no. 12: 1942–51.

Gervais, Will. 2017. "Post-Publication Peer Review." *Will Gervais*, 3 February. http://willgervais.com/blog/2017/3/2/post-publication-peer-review.

Giambra, Leonard M. 1995. "A Laboratory Method for Investigating Influences on Switching Attention to Task-Unrelated Imagery and Thought." *Consciousness and Cognition* 4, no. 1: 1–21.

Gibbs, Frederic A., and Erna L. Gibbs. 1941. *Atlas of Encephalography*. Reading, MA: Addison-Wesley.

Gilbert, Daniel T., Gary King, Stephen Pettigrew, and Timothy D. Wilson. 2016. "Comment on 'Estimating the Reproducibility of Psychological Science." *Science* 351, no. 6277: 1037. https://doi.org/10.1126/science.aad7243.

Gillespie, Tarleton. 2016. "Algorithm." In *Digital Keywords: A Vocabulary of Information Society and Culture*, ed. Benjamin Peters, 18–30. Princeton, NJ: Princeton University Press.

Giordano, James J., and Bert Gordijn, eds. 2010. *Scientific and Philosophical Perspectives in Neuroethics*. Cambridge, UK: Cambridge University Press.

Gitelman, Lisa, and Virginia Jackson. 2013. "Introduction." In *"Raw Data" Is an Oxymoron*, ed. Lisa Gitelman, 1–14. Cambridge, MA: MIT Press.

Goldstine, Hermann Heine, and John von Neumann. 1947. *Planning and Coding of Problems for an Electric Computing Instrument*. Princeton, NJ: Institute for Advanced Studies.

Gomez-Marin, Alex, Joseph J. Patton, Adam R. Kamof, Rui M. Costa, and Zachary F. Mainen. 2014. "Big Behavioural Data: Psychology, Ethology and the Foundations of Neuroscience." *Nature Neuroscience* 17, no. 11: 1455–62.

Gonzalez-Castillo, Javier, Ziad S. Saad, Daniel A. Handwerker, Souheil J. Inati, Noah Brenowitz, and Peter A. Bandettini. 2012. "Whole-Brain, Time-Locked Activation with Simple Tasks Revealed Using Massive Averaging and Model-Free Analysis." *Proceedings of the National Academy of Sciences* 109, no. 14: 5487–92.

Gould, Paula. 2004. "Physicists Help to Decode the Brain." *Physics World* 17, no. 4: 12.

Gramelsberger, Gabriele, ed. 2011. *From Science to Computational Sciences: Studies in the History of Computing and Its Influence on Today's Sciences.* Zürich: Diaphanes.

Greicius, Michael D., Ben Krasnow, Allan L. Reiss, and Vinod Menon. 2003. "Functional Connectivity in the Resting Brain: A Network Analysis of the Default Mode Hypothesis." *Proceedings of the National Academy of Sciences* 100, no. 1: 253–8.

Greicius, Michael D., and Vinod Menon. 2004. "Default-Mode Activity during a Passive Sensory Task: Uncoupled from Deactivation but Impacting Activation." *Journal of Cognitive Neuroscience* 16, no. 9: 1484–92.

Gruber, David, and Jacob A. Dickerson. 2012. "Persuasive Images in Popular Science: Testing Judgments of Scientific Reasoning and Credibility." *Public Understanding of Science* 21, no. 8: 938–48.

Gu, Shi, Matthew Cislak, Benjamin Baird, Sarah F. Muldoon, Scott T. Grafton, Fabio Pasqualetti, and Danielle Bassett. 2018. "The Energy Landscape of Neurophysiological Activity Implicit in Brain Network Structure." *Nature Scientific Reports* 8, art. 2507: 1–15. https://doi.org/10.1038/s41598-018-20123-8.

Gusnard, Debra A., and Marcus E. Raichle. 2001. "Searching for a Baseline: Functional Imaging and the Resting Human Brain." *Nature Reviews Neuroscience* 2, no. 10: 685–94.

Gusnard, Debra A., Erbil Akbudak, Gordon L. Shulman, and Marcus E. Raichle. 2001. "Medial Prefrontal Cortex and Self-Referential Mental Activity: Relation to a Default Mode of Brain Function." *Proceedings of the National Academy of Sciences* 98, no. 7: 4259–64.

Hagner, Michael, and Cornelius Borck. 2001. "Mindful Practices: On the Neurosciences in the Twentieth Century." *Science in Context* 14, no. 4: 507–10.

Haigh, Thomas. 2008. "Cleve Moler: Mathematical Software Pioneer and Creator of Matlab." IEEE *Annals of the History of Computing* 30, no. 1: 87–91.

Haigh, Thomas, Mark Priestly, and Crispin Hope. 2016. ENIAC *in Action: Making and Remaking the Modern Computer.* Cambridge, MA: MIT Press.

Halpern, Orit, Jesse Lecavalier, Nerea Calvillo, and Wolfgang Pietsch. 2013. "Test-Bed Urbanism." *Public Culture* 25, no. 2: 272–306.

Hanrahan, Brían. 2015. "The Anthropoid Condition: Brían Hanrahan

Interviews John Durham Peters." *Los Angeles Review of Books*, 10 July. https://lareviewofbooks.org/article/the-anthropoid-condition-an-interview-with-john-durham-peters.

Haraway, Donna. 1988. "Situated Knowledges: The Science Question in Feminism and the Privilege of Partial Perspective." *Feminist Studies* 14, no. 3: 575–99. https://philpapers.org/archive/HARSKT.pdf.

Harcourt, Bernard E. 2007. *Against Prediction: Profiling, Policing, and Punishing in an Actuarial Age*. Chicago: University of Chicago Press.

Hartgerink, Chris H.J. 2017. "Reanalyzing Head et al. (2015): Investigating the Robustness of Widespread P-Hacking." *PeerJ* 5, no. 3068: 1–10. https://doi.org/10.7717/peerj.3068.

Hassabis, Demis, Darshan Kumaran, Christopher Summerfield, and Matthew Botvinick. 2017. "Neuroscience-Inspired Artificial Intelligence." *Neuron* 95, no. 2: 245–58.

Hassabis, Demis, Dharshan Kuraman, Seralynne D. Vann, and Eleanor A. Maguire. 2007. "Patients with Hippocampal Amnesia Cannot Imagine New Experiences." *Proceedings of the National Academy of Sciences* 104, no. 5: 1726–31.

Hassabis, Demis, R. Nathan Spreng, Andrei A. Rusu, Clifford A. Robbins, Raymond A. Mar, and Daniel L. Schacter. 2014. "Imagine All the People: How the Brain Creates and Uses Personality Models to Predict Behavior." *Cerebral Cortex* 24, no. 8: 1979–87. https://doi.org/10.1093/cercor/bht042.

Haugeland, John. 1985. *Artificial Intelligence: The Very Idea*. Reprint, Cambridge, MA: MIT Press, 1989.

Hausken, Liv, Bettina Papenburg, and Sigrid Schmitz. 2018. "Introduction: The Processes of Imaging / The Imaging of Processes." *Catalyst: Feminism, Theory, Technoscience* 4, no. 2: 1–23.

Head, Meghan, Luke Holman, Rob Lanfear, Andrew T. Kahn, and Michael D. Jennions. 2015. "The Extent and Consequences of P-Hacking in Science." *PLOS Biology* 13, no. 3: 1–15. https://doi.org/10.1371/journal.pbio.1002106.

Hecht, Gabrielle. 2018. "Interscalar Vehicles for an African Anthropocene: On Waste, Temporality, and Violence." *Cultural Anthropology* 33, no. 1: 109–41.

Helmreich, Stefan. 2000. *Silicon Second Nature: Culturing Artificial Life in a Digital World*. Berkeley: University of California Press.

Hinton, Geoffrey, Simon Osindero, and Yee-Whye Teh. 2006. "A Fast Learning Algorithm for Deep Belief Nets." *Neural Computation* 18, no. 7: 1527–54.

Hollin, Gregory. 2017. "Autistic Heterogeneity: Linking Uncertainties and Indeterminacies." *Science as Culture* 26, no. 2: 209–31.

Hollin, Gregory, and Alison Pilnick. 2015. "Infancy, Autism, and the Emergence of a Socially Disordered Body." *Social Science and Medicine* 143: 279–86.

Hook, Cayce J., and Martha J. Farah. 2013. "Look Again: Effects of Brain Images and Mind-Brain Dualism on Lay Evaluations of Research." *Journal of Cognitive Neuroscience* 25, no. 9: 1397–405.

Hopkins, Emily J., Deena S. Weisberg, and Jordan C.V. Taylor. 2016. "The Seductive Allure Is a Reductive Allure: People Prefer Scientific Explanations that Contain Logically Irrelevant Reductive Information." *Cognition* 155: 67–76.

Hu, Tung-Hui. 2015. *A Prehistory of the Cloud*. Cambridge, MA: MIT Press.

– 2017. "Black Boxes and Green Lights: Media, Infrastructure, and the Future at Any Cost." *English Language Notes* 55, nos 1–2: 81–8.

Illes, Judy, and Stephanie J. Bird. 2006. "Neuroethics: A Modern Context for Ethics in Neuroscience" *Trends in Neurosciences* 29, no. 9: 511–17.

Ioannidis, John P.A. 2005. "Why Most Published Research Findings Are False." *PLOS Medicine* 2, no. 8: 0696–0701. https://doi.org/10.1371/journal.pmed.0020124.

– 2012. "Why Science Is Not Necessarily Self-Correcting." *Perspectives on Psychological Science* 7, no. 6: 645–54.

Jabbi, Mbemba, Christian Keysers, Tania Singer, and Klaas Enno Stephan. 2009. "Response to 'Voodoo Correlations in Social Neuroscience' by Vul et al. – Summary Information for the Press." http://cogns.northwestern.edu/cbmg/replyVul.pdf.

Jarrett, Christian. 2012. "Mirror Neurons: The Most Hyped Concept in Neuroscience." *Psychology Today*, 10 December. http://www.psychologytoday.com/blog/brain-myths/201212/mirror-neurons-the-most-hyped-concept-in-neuroscience.

Jensen, Casper Bruun, and Atsuro Morita. 2015. "Infrastructures as Ontological Experiments." *Engaging Science, Technology, and Society* 1: 81–7.

Jiahui, Sun. 2017. "The Alpha and Omega of Go." *The World of Chinese*, 20 July. http://www.theworldofchinese.com/2017/07/the-alpha-and-omega-of-go.

Jonas, Eric, and Konrad P. Kording. 2017. "Could a Neuroscientist Understand a Microprocessor?" *PLOS Computational Biology* 13, no. 1: 1–24. https://doi.org/10.1371/journal.pcbi.1005268.

joneilortiz. 2008. "Vul on fMRI Abuse in the Cognitive Neuroscience of Social Interaction." *Mutually Occluded: Media and Film, Design, Philosophy, Politics*, 30 December.

Jones, Owen D., René Marois, Martha J. Farah, and Henry T. Greely. 2013. "Law and Neuroscience." *Journal of Neuroscience* 33, no. 45: 17624–30.

Joyce, Kelly A. 2008. *Magnetic Appeal: MRI and the Myth of Transparency*. Ithaca, NY: Cornell University Press.

Jung, Rex E., and Oshin Vartanian, eds. 2018. *The Cambridge Handbook of the Neuroscience of Creativity*. Cambridge, UK: Cambridge University Press.

Keller, Evelyn Fox. 2003. *Making Sense of Life: Explaining Biological Development with Models, Metaphors, and Machines*. Cambridge, MA: Harvard University Press.

Kelly, Éanna. 2015. "€1B Human Brain Project Back on Track after Commission Signs New Contract." *Science/Business*, 3 November. https://sciencebusiness.net/news/77294/%E2%82%AC1B-Human-Brain-Project-back-on-track-after-Commission-signs-new-contract.

Kennedy, Charles B. 1991. "Louis Sokoloff at Three Score and Ten." *Journal of Cerebral Blood Flow and Metabolism* 11, no. 6: 885–9.

Kevles, Bettyann H. 1997. *Naked to the Bone: Medical Imaging in the Twentieth Century*. New Brunswick, NJ: Rutgers University Press.

Kietzmann, Tim C., Patrick McClure, and Nikolaus Kriegeskorte. 2019. "Deep Neural Networks in Computational Neuroscience." *Oxford Research Encyclopedia: Neuroscience*, 1–28. https://oxfordre.com/neuroscience/abstract/10.1093/acrefore/9780190264086.001.0001/acrefore-9780190264086-e-46.

Kilner, J.M., and R.N. Lemon. 2013. "What We Know Currently about Mirror Neurons." *Current Biology* 23, no. 23: 1057–62.

Kirkpatrick, James, Razvan Pascanu, Neil Rabinowitz, Joel Veness, Guillaume Desjardins, Andrei A. Rusu, Kieran Milan, et al. 2017. "Overcoming Catastrophic Forgetting in Neural Networks." *Proceedings of the National Academy of Sciences* 114, no. 13: 3521–6.

Kitchin, Rob. 2017. "Thinking Critically about and Researching Algorithms." *Information, Communication and Society* 20, no. 1: 14–29.

Klein, Julie T. 1994. *Interdisciplinarity: History, Theory, and Practice*. Detroit: Wayne State University Press.

Knuuttila, Tarja, and Andrea Loettgers. 2014a. "Magnets, Spins, and Neurons: The Dissemination of Model Templates across Disciplines." *The Monist* 97, no. 3: 280–300.

– 2014b. "Varieties of Noise: Analogical Reasoning in Synthetic Biology." *Studies in History and Philosophy of Science Part A* 48: 76–88.

– 2016. "Modelling as Indirect Representation? The Lotka-Volterra Model Revisited." *British Journal for the Philosophy of Science* 68, no. 4: 1007–36.

Kriegeskorte, Nikolaus, and Pamela K. Douglas. 2018. "Cognitive Computational Neuroscience." *arXiv*, 31 July, 1–31. https://arxiv.org/abs/1807.11819.

Kriegeskorte, Nikolaus, W. Kyle Simmons, Patrick S.F. Bellgowan, and Chris I. Baker. 2009. "Circular Analysis in Systems Neuroscience: The Dangers of Double Dipping." *Nature Neuroscience* 12, no. 5: 535–40.

Landau, William M., Walter H. Freygang Jr, Lewis P. Roland, Louis Sokoloff, and Seymour Kety. 1955. "The Local Circulation of the Living Brain: Values in the Unanesthetized and Anesthetized Cat." *Transactions of the American Neurological Association* 80: 125–9. https://profiles.nlm.nih.gov/ps/access/NLBBCK.pdf.

Larkin, Brian. 2013. "The Politics and Poetics of Infrastructure." *Annual Review of Anthropology* 42: 327–43.

– 2015. "Form." *Cultural Anthropology*, 24 September. https://culanth.org/fieldsights/form.

– 2016. "Ambient Infrastructures: Generator Life in Nigeria." *Technosphere Magazine*, 15 November. https://technosphere-magazine.hkw.de/p/Ambient-Infrastructures-Generator-Life-in-Nigeria-fCgtKng7vpt7otmky9vnFw.

Latour, Bruno. 1987. *Science in Action: How to Follow Scientists and Engineers through Society*. Cambridge, MA: Harvard University Press.

– 1999. *Pandora's Hope: Essays on the Reality of Science Studies*. Cambridge, MA: Harvard University Press.

– 2012. "Visualisation and Cognition: Drawing Things Together." *Avant: Trends in Interdisciplinary Studies* 3: 207–60.

Lehrer, Jonathan. 2008. "Can a Thinking, Remembering, Decision-Making, Biologically Accurate Brain Be Built from a Supercomputer? Out of the Blue" *Seed*, 3 March. https://www.douban.com/group/topic/8956401.

– 2009. "Voodoo Correlations: Have the Results of Some Brain Scanning Experiments Been Overstated?" *Scientific American*, 29 January. http://www.scientificamerican.com/article.cfm?id=brain-scan-results-overstated.

Leibo, Joel Z., Cyprien de Masson d'Autume, Daniel Zoran, David Amos, Charles Beattie, Keith Anderson, Antonio García Castañeda, et al. 2018.

"Psychlab: A Psychology Laboratory for Deep Reinforcement Learning Agents." *arXiv*, 4 February, 1–28. https://arxiv.org/abs/1801.08116.

Lemenager, Stephanie. 2016. "Infrastructure Again, and Always." *Reviews in Cultural Theory* 6, no. 1: 25–9. http://reviewsinculture.com/2016/03/09/infrastructure-again-and-always.

Lemov, Rebecca. 2010. "'Hypothetical Machines': The Science Fiction Dreams of Cold War Social Science." *Isis* 101, no. 2: 401–11.

Leonelli, Sabina, Brain Rappert, and Gail Davies. 2017. "Data Shadows: Knowledge, Openness, and Absence." *Science, Technology, and Human Values* 42, no. 2: 191–202.

Levin, Nadine. 2014. "Multivariate Statistics and the Enactment of Metabolic Complexity." *Social Studies of Science* 44, no. 4: 555–78.

Lieberman, Matthew D., Elliot T. Berkman, and Tor D. Wager. 2009. "Correlations in Social Neuroscience Aren't Voodoo: Commentary on Vul et al. (2009)." *Perspectives on Psychological Science* 4, no. 3: 299–307.

Lingnau, Angelika, Benno Gesierich, and Alfonso Caramazza. 2009. "Asymmetric fMRI Adaptation Reveals No Evidence for Mirror Neurons in Humans." *Proceedings of the National Academy of Sciences* 106, no. 24: 9925–30.

Llinás, Rodolfo. 2001. *I of the Vortex: From Neurons to Self*. Cambridge, MA: MIT Press.

Logothetis, Nikos K., and Josef Pfeuffer. 2004. "On the Nature of the BOLD fMRI Contrast Mechanism." *Magnetic Resonance Imaging* 22, no. 10: 1517–31.

Lohmann, Gabriele, Johannes Stelzer, Karsten Müller, Eric Lacosse, Tilo Buschmann, Vinod Kumar, Wolfgang Grodd, and Klaus Scheffler. 2017. "Inflated False Negative Rates Undermine Reproducibility in Task-Based MRI." *bioRxiv*, 31 March, 1–18. https://www.biorxiv.org/content/10.1101/122788v1.

Love, Bradley C. 2015. "The Algorithmic Level Is the Bridge between Computation and Brain." *Topics in Cognitive Science* 7, no. 2: 230–42.

Lowrie, Ian. 2017. "Algorithmic Rationality: Epistemology and Efficiency in the Data Sciences." *Big Data and Society* 4, no. 1: 1–13. https://doi.org/10.1177/2053951717700925.

Luck, Steve. 2018. "Why I've Lost Faith in P Values." *Luck Lab*, 18 April. https://lucklab.ucdavis.edu/blog/2018/4/19/why-i-lost-faith-in-p-values.

Lussier, Kira. 2018. "From the Intuitive Human to the Intuitive Computer." *Technology's Stories*, 12 March. http://www.technologystories.org/intuitive-computer.

Lyon, Louisa. 2017. "Dead Salmon and Voodoo Correlations: Should We Be Sceptical about Functional MRI?" *Brain: A Journal of Neurology* 140, no. 8, e53: 1–5. https://doi.org/10.1093/brain/awx180.

Machamer, Peter K., Marcello Pera, and Aristeidēs Baltas, eds. 2000. *Scientific Controversies: Philosophical and Historical Perspectives.* New York: Oxford University Press.

Mackenzie, Adrian. 2017. *Machine Learners: Archaeology of a Data Practice.* Cambridge, MA: MIT Press.

MacLeod, Miles, and Nancy Nersessian. 2018. "Modelling Complexity: Cognitive Constraints and Computational Model-Building in Integrative Systems Biology." *History and Philosophy of the Life Sciences* 40, no. 17: 1–28. https://doi.org/10.1007/s40656-017-0183-9.

Majaca, Antonia, and Luciana Parisi. 2016. "The Incomputable and Instrumental Possibility." *e-flux*, no. 77. https://www.e-flux.com/journal/77/76322/the-incomputable-and-instrumental-possibility.

Malinowska, Joanna K. 2016. "Cultural Neuroscience and the Category of Race: The Case of the Other-Race Effect." *Synthese* 193, no. 12: 3865–87.

Marblestone, Adam H., Greg Wayne, and Konrad P. Kording. 2016. "Toward an Integration of Deep Learning and Neuroscience." *Frontiers in Computational Neuroscience* 10, art. 94: 1–41. https://doi.org/10.3389/fncom.2016.00094.

Marchetti, Antonella, Francesca Baglio, Isa Costantini, Ottavia Dipasquale, Federica Savazzi, Raffaello Nemni, Francesca Sangiuliano Intra, et al. 2015. "Theory of Mind and the Whole Brain Functional Connectivity: Behavioral and Neural Evidences with the Amsterdam Resting State Questionnaire." *Frontiers in Psychology* 6, art. 1855: 1–10. https://www.frontiersin.org/articles/10.3389/fpsyg.2015.01855/full.

Margulies, Daniel. 2011. "The Salmon of Doubt: Six Months of Methodological Controversy within Social Neuroscience." In *Critical Neuroscience: A Handbook of the Social and Cultural Contexts of Neuroscience*, ed. Soupharna Choudhury and Jan Slaby, 273–85. Oxford: Wiley-Blackwell.

Markram, Henry, Eilif Muller, Srikanth Ramaswamy, Michael W. Reimann, Marwan Abdellah, Carlos Aguado Sanchez, Anastasia Ailamaki, et al. 2015. "Reconstruction and Simulation of Neocortical Microcircuitry." *Cell* 163, no. 2: 456–92.

Martin, Emily. 2010. "Self-Making and the Brain." *Subjectivity* 3, no. 4: 366–81.

Martinez Mateo, M., M. Cabanis, J. Stenmans, and S. Krach. 2013. "Essentializing the Binary Self: Individualism and Collectivism in Cultural Neuroscience." *Frontiers in Human Neuroscience* 7, art. 289: 1–4. https://doi.org/10.3389/fnhum.2013.00289.

Mason, M.F., M.I. Norton, J.D. van Horn, D.M. Wegner, S.T. Grafton, and C.N. Macrae. 2007. "Wandering Minds: The Default Network and Stimulus-Independent Thought." *Science* 315, no. 5810: 393–5.

Mattern, Shannon. 2013. "Infrastructural Tourism." *Places Journal*, July. https://doi.org/10.22269/130701.

– 2016. "Cloud and Field." *Places Journal*, August. https://doi.org/10.22269/160802.

– 2018. "Scaffolding, Hard and Soft: Critical and Generative Infrastructures." In *The Routledge Companion to Media Studies and Digital Humanities*, ed. Jentery Sayers, 318–26. New York: Routledge.

Mattila, Erika. 2005. "Interdisciplinarity 'In the Making': Modelling Infectious Diseases." *Perspectives on Science* 13, no. 4: 531–54.

Matusall, Svenja. 2012. "Searching for the Social in the Brain: The Emergence of Social Neuroscience." PhD diss., Eidgenössische Technische Hochschule Zurich.

– 2013. "Social Behavior in the 'Age of Empathy'? – A Social Scientist's Perspective on Current Trends in the Behavioral Sciences." *Frontiers in Human Neuroscience* 7, art. 236: 1–5. https://doi.org/10.3389/fnhum.2013.00236.

Matusall, Svenja, Ina Kaufmann, and Markus Christen. 2011. "The Emergence of Social Neuroscience as an Academic Discipline." In *The Oxford Handbook of Social Neuroscience*, ed. John Decety and John Cacioppo, 9–27. Oxford: Oxford University Press.

Mazoyer, Bernard, Laure Zago, Emanuel Mellet, Stephanie Bricogne, Olivier Etard, Olivier Houdé, Fabrice Crivello, et al. 2001. "Cortical Networks for Working Memory and Executive Functions Sustain the Conscious Resting State in Man." *Brain Research Bulletin* 54, no. 3: 287–98.

Mazziotta, John, Michael E. Phelps, Richard E. Carson, and David E. Kuhl. 1982. "Tomographic Mapping of Human Cerebral Metabolism: Sensory Deprivation." *Annals of Neurology* 12, no. 5: 435–44.

Mazziotta, John, Arthur Toga, Alan Evans, Peter Fox, Jack Lancaster, Karl Zilles, Roger Woods, et al. 2001. "A Probabilistic Atlas and Reference System for the Human Brain: International Consortium for Brain Mapping (ICBM)." *Philosophical Transactions of the Royal Society B Biological Sciences* 356, no. 1412: 1293–1322.

Mazziotta, John, Roger Woods, Marco Iacoboni, Nancy Sicotte, Kami Yaden, Mary Tran, Courtney Bean, Jonas Kaplan, and Arthur Toga. 2009. "The Myth of the Normal, Average Human Brain – The ICBM Experience: (1) Subject Screening and Eligibility." *NeuroImage* 44, no. 3: 914–22.

McCabe, David P., and Alan D. Castel. 2008. "Seeing Is Believing: The Effect of Brain Images on Judgments of Scientific Reasoning." *Cognition* 107, no. 1: 343–52.

McCarthy, J. 1990. "Chess as the Drosophila of AI." In *Computers, Chess, and Cognition*, ed. T. Anthony Marsland and Jonathan Schaeffer, 227–38. New York: Springer-Verlag.

McCulloch, Warren S. 1945. "A Heterarchy of Values Determined by the Topology of Nervous Nets." *Bulletin of Mathematical Biophysics* 7, no. 2: 89–93.

McCulloch, Warren S., and Walter H. Pitts. 1988. "A Logical Calculus of Ideas Immanent in Nervous Activity." In *Embodiments of Mind*, ed. Warren S. McCulloch, 19–39. Cambridge, MA: MIT Press.

Mendelsohn, Ben, and Alex Chohlas-Wood. 2011. "Bundled, Buried and behind Closed Doors." *Vimeo.* https://vimeo.com/30642376.

Menon, Vinod. 2011. "Large-Scale Brain Networks and Psychopathology: A Unifying Triple Network Model." *Trends in Cognitive Science* 15, no. 10: 483–506.

Merolla, Paul A., John V. Arthur, Rodrigo Alvarez-Icaza, Andrew S. Cassidy, Jun Sawada, Filipp Akopyan, Bryan L. Jackson, et al. 2014. "A Million Spiking-Neuron Integrated Circuit with a Scalable Communication Network and Interface." *Science* 345, no. 6197: 668–73. https://doi.org/10.1126/science.1254642.

Merritt, David. 2017. "Cosmology and Convention." *Studies in History and Philosophy of Science Part B: Studies in History and Philosophy of Modern Physics* 57: 41–52.

Merton, Robert K., Marjorie Fiske, and Patricia Kendall. 1956. "The Focussed Interview." *American Journal of Sociology* 51: 541–57.

Merz, Martina. 1998. "'Nobody Can Force You When You Are across the Ocean' – Face to Face and E-Mail Exchanges between Theoretical Physicists." In *Making Space for Science: Territorial Themes in the Shaping of Knowledge*, ed. C. Smith, J. Agar, and G. Schmidt, 313–39. New York: St Martin's.

– 1999. "Multiplex and Unfolding: Computer Simulation in Particle Physics" *Science in Context* 12, no. 2: 293–316.

– 2006. "Embedding Digital Infrastructure in Epistemic Culture." In *New*

Infrastructures for Knowledge Production: Understanding E-Science, ed. C. Hine, 99–119. Hershey, PA: Information Science Publishing.

Messeri, Lisa. 2016. *Placing Outer Space: An Earthly Ethnography of Other Worlds*. Durham, NC: Duke University Press.

Metz, Cade. 2016a. "In Two Moves, AlphaGo and Lee Sedol Redefined the Future." *Wired*, 16 March. https://www.wired.com/2016/03/two-moves-alphago-lee-sedol-redefined-future.

– 2016b. "The Sadness and Beauty of Watching Google's AI Play Go." *Wired*, 11 March. https://www.wired.com/2016/03/sadness-beauty-watching-googles-ai-play-go.

Michael, Robert B., Eryn J. Newman, Matti Vuorre, Geoff Cumming, and Maryanne Garry. 2013. "On the (Non)persuasive Power of a Brain Image." *Psychonomic Bulletin and Review* 20, no. 4: 720–5.

Morawski, Jill. 2007. "Scientific Selves: Discerning the Subject and the Experimenter in Experimental Psychology in the United States, 1900–1935." In *Psychology's Territories: Historical and Contemporary Perspectives from Different Disciplines*, ed. Mitchell G. Ash and Thomas Sturm, 129–48. Mahwah and London: Lawrence Erlbaum Associates.

– 2015. "Epistemological Dizziness in the Psychology Laboratory: Lively Subjects, Anxious Experimenters, and Experimental Relations, 1950–1970." *Isis* 106, no. 3: 567–97.

Morcom, Alexa M., and Paul C. Fletcher. 2007. "Does the Brain Have a Baseline? Why We Should Be Resisting Rest." *NeuroImage* 37, no. 4: 1073–82.

Morgan, Mary S. 2004. "Simulation: The Birth of a Technology to Create 'Evidence' in Economics." *Revue d'histoire des Sciences* 57, no. 2: 339–75.

Mueller, Karsten, Jöran Lepsien, Harald E. Möller, and Gabriele Lohmann. 2017. "Commentary: Cluster Failure: Why fMRI Inferences for Spatial Extent Have Inflated False-Positive Rates." *Frontiers in Human Neuroscience* 11, art. 345: 1–3. https://doi.org/10.3389/fnhum.2017.00345.

Muller, Eilif, James Bednar, Markus Diesmann, Marc-Oliver Gewaltig, Michael Hines, and Andrew Davison. 2015. "Python in Neuroscience." *Frontiers in Neuroinformatics* 9, art. 11: 1–4. https://doi.org/10.3389/fninf.2015.00011.

Munster, Anna. 2013. *An Aesthesia of Networks: Conjunctive Experience in Art and Technology*. Cambridge, MA: MIT Press.

Myers, Greg. 1990. *Writing Biology: Texts in the Social Construction of Scientific Knowledge*. Madison: University of Wisconsin Press.

Myers, Natasha. 2015. *Rendering Life Molecular: Models, Modelers, and Excitable Matter*. Durham, NC: Duke University Press.

Nadesan, Majia Holmer. 2005. *Constructing Autism: Unravelling the 'Truth' and Understanding the Social*. London: Routledge.

Nelkin, Dorothy. 1992. *Controversy: Politics of Technical Decisions*. Newbury Park, CA: Sage.

Nelson, Leif, Joseph Simmons, and Uri Simonsohn. 2018. "Psychology's Renaissance." *Annual Review of Psychology* 69: 511–34.

Neurocritic. 2008. "Scan Scandal Hits Social Neuroscience." *The Neurocritic*, 31 December. https://neurocritic.blogspot.com/2008/12/scan-scandal-hits-social-neuroscience.html.

– 2012. "The Not So Seductive Allure of Colorful Brain Images." *The Neurocritic*, 7 December. https://neurocritic.blogspot.com/2012/12/the-not-so-seductive-allure-of-colorful_7.html.

Neuroskeptic. 2009. "'Voodoo Correlations' in fMRI – Whose Voodoo?" *Discover*, 4 February. http://blogs.discovermagazine.com/neuroskeptic/2009/02/04/voodoo-correlations-in-fmri-whose-voodoo/#.XGhs8c9KjOQ.

Nichols, Thomas E., and Jean-Baptist Poline. 2009. "Commentary on Vul et al.'s (2009) 'Puzzlingly High Correlations in fMRI Studies of Emotion, Personality, and Social Cognition.'" *Perspectives on Psychological Science* 4, no. 3: 291–3.

Noble, Safiya Umoja. 2018. *Algorithms of Oppression: How Search Engines Reinforce Racism*. New York: New York University Press.

Oaten, Megan, Kipling D. Williams, Andrew Jones, and Lisa Zadro. 2008. "The Effects of Ostracism on Self-Regulation in the Socially Anxious." *Journal of Social and Clinical Psychology* 27, no. 5: 471–504.

Ogawa, Seiji, Tso-Ming Lee, A.R. Kay, and D.W. Tank. 1990. "Brain Magnetic Resonance Imaging with Contrast Dependent on Blood Oxygenation." *Proceedings of the National Academy of Science* 87, no. 24: 9868–72.

Olson, Valerie. 2018. *Into the Extreme: U.S. Environmental Systems and Politics beyond Earth*. Minneapolis: University of Minnesota Press.

Open Science Collaboration. 2015. "Estimating the Reproducibility of Psychological Science." *Science* 349, no. 6251: 943–54. https://doi.org/10.1126/science.aac4716.

Ortega, Francisco, and Fernando Vidal. 2007. "Mapping the Cerebral Subject in Contemporary Culture." *RECIIS* 1, no. 2: 255–9.

– eds. 2011. *Neurocultures: Glimpses into an Expanding Universe*. Frankfurt am Main: Peter Lang.

Özden-Schilling, Canay. 2015. "Economy Electric." *Cultural Anthropology* 30, no. 4: 578–88.

– 2016. "The Infrastructure of Markets: From Electric Power to Electronic Data." *Economic Anthropology* 3, no. 1: 68–80.

Padmanabhan, Aarthi, Charles J. Lynch, Marie Schaer, and Vinod Menon. 2017. "The Default Mode Network in Autism." *Biological Psychiatry: Cognitive Neuroscience and Neuroimaging* 2, no. 6: 476–86. https://doi. org/10.1016/j.bpsc.2017.04.004.

Parks, Lisa. and Nicole Starosielski, eds. 2015. *Signal Traffic: Cultural Studies of Media Infrastructures*. Champaign: University of Illinois Press.

Pasquale, Frank. 2015. *The Black Box Society*. Cambridge, UK: Cambridge University Press.

Pasquinelli, Matteo. 2015. "What an Apparatus Is Not: On the Archaeology of the Norm in Foucault, Canguilhem and Goldstein." *Parrhesia*, no. 22: 79–89.

Perez, Carlos E. 2017. "How Cargo Cult Bayesians Encourage Deep Learning Alchemy." *Medium*, 16 November. https://medium.com/ intuitionmachine/cargo-cult-statistics-versus-deep-learning-alchemy-8d7700134c8e.

Perkel, Jeffrey M. 2015. "Programming: Pick up Python." *Nature* 518: 125–6. https://www.nature.com/news/programming-pick-up-python-1.16833.

Peters, Tim. 2004. "The Zen of Python." *GitHub*, 22 August. https:// github.com/python/peps/blob/master/pep-0020.txt.

Pias, Claus. 2011. "On the Epistemology of Computer Simulation." *Zeitschrift für Medien- und Kulturforschung*, no. 1: 29–54.

Pickersgill, Martyn, and Ira van Keulen, eds. 2011. *Advances in Medical Sociology*. Vol. 13, *Sociological Reflections on the Neurosciences*. Bingley, UK: Emerald.

Pinch, Trevor, and Wiebe Bijker. 1984. "The Social Construction of Facts and Artefacts: Or How the Sociology of Science and the Sociology of Technology Might Benefit Each Other." *Social Studies of Science* 14, no. 3: 399–441.

Pitts, Walter H., and Warren S. McCulloch. 1988. "How We Know Universals: The Perception of Auditory and Visual Forms." In *Embodiments of Mind*, ed. Warren S. McCulloch, 46–66. Cambridge, MA: MIT Press.

Pitts-Taylor, Victoria. 2014. "Cautionary Notes on Navigating the Neurocognitive Turn." *Sociological Forum* 29, no. 4: 995–1000.

Poerio, Giulia L., and Jonathan Smallwood. 2016. "Daydreaming to Navigate the Social World: What We Know, What We Don't Know, and Why It Matters." *Social and Personality Psychology Compass* 10, no. 11: 605–18.

Poldrack, Russell A., and Jeanette A. Mumford. 2009. "Independence in ROI Analysis: Where Is the Voodoo?" *Social Cognitive and Affective Neuroscience* 4, no. 2: 208–13.

Posner, Miriam. 2018. "See No Evil." *Logic* 4: 215–29. https://logicmag.io/04-see-no-evil.

Powell, Hilary, Hazel Morrison, and Felicity Callard. 2018. "Wandering Minds: Tracing Inner Worlds through a Historical-Geographical Art Installation." *GeoHumanities* 4, no. 1: 132–56.

Prasad, Amit. 2007. "The (Amorphous) Anatomy of an Invention: The Case of Magnetic Resonance Imaging (MRI)." *Social Studies of Science* 37, no. 4: 533–60.

Pritzel, Alexander, Benigno Uria, Sriram Srinivasan, Adrià Puigdomènech, Oriol Vinylas, Demis Hassabis, Daan Wierstra, and Charles Blundell. 2017. "Neural Episodic Control." *arXiv*, 6 March, 1–12. https://arxiv.org/abs/1703.01988.

Quora. 2014. "What Are the Main Objections to the Human Brain Project?" https://www.quora.com/What-are-the-main-objections-to-the-human-brain-project.

Racine, Eric, and Zoe Costa-von Aesch. 2011. "Neuroscience's Impact on Our Self-Identity: Perspectives from Ethics and Public Understanding." In *Neurocultures: Glimpses into an Expanding Universe*, ed. Francisco Ortega and Fernando Vidal, 83–98. Frankfurt am Main: Peter Lang.

Racine, Eric, and Emma Zimmerman. 2012. "Pragmatic Neuroethics and Neuroscience's Potential to Radically Change Ethics." In *The Neuroscientific Turn: Transdisciplinarity in the Age of the Brain*, ed. Melissa M. Littlefield and Jenelle M. Johnson, 135–51. Ann Arbor: University of Michigan Press.

Raichle, Marcus. 2006. "The Brain's Dark Energy." *Science* 314, no. 5803: 1249–50.

– 2010. "Two Views of Brain Function." *Trends in Cognitive Sciences* 14, no. 4: 180–90.

– 2015a. "The Brain's Default Mode Network." *Annual Review of Neuroscience* 38: 433–47. https://doi.org/10.1146/annurev-neuro-071013-014030.

– 2015b. "The Restless Brain: How Intrinsic Activity Organizes Brain Function." *Philosophical Transactions of The Royal Society B Biological Sciences* 370, no. 1668: 1–11.

Raichle, Marcus E., Ann Mary McLeod, Abraham Z. Snyder, William J. Powers, Debra A. Gusnard, and Gordon L. Shulman. 2001. "A Default Mode of Brain Function." *Proceedings of the National Academy of Sciences* 98, no. 2: 676–82.

Rees, Tobias. 2018. *After Ethnos*. Durham, NC: Duke University Press.

Rieder, Bernhard. 2017. "Scrutinizing an Algorithmic Technique: The Bayes Classifier as Interested Reading of Reality." *Information, Communication and Society* 20, no. 1: 100–17.

Ritter, Samuel, Jane X. Wang, Zeb Kurth-Nelson, and Matthew Botvinick. 2018. "Episodic Control as Meta-Reinforcement Learning." *bioRxiv*, 6 July, 1–7. https://www.biorxiv.org/content/10.1101/360537v2.

Rodgers, Adam. 2017. "Star Neuroscientist Tom Insel Leaves the Google-Spawned Verily for … a Startup?" *Wired*, 11 May. https://www.wired.com/2017/05/star-neuroscientist-tom-insel-leaves-google-spawned-verily-startup.

Roepstorff, Andreas. 2001. "Brains in Scanners: An Umwelt of Cognitive Neuroscience." *Semiotica* 134: 747–65.

– 2002. "Transforming Subjects into Objectivity: An 'Ethnography of Knowledge.'" *Folk: Journal of the Danish Ethnographic Society* 44: 145–70.

– 2013. "Why I Am Not Just Lovin' Cultural Neuroscience? Toward a Slow Science of Cultural Difference." *Psychological Inquiry: An International Journal for the Advancement of Psychological Theory* 24, no. 1: 61–3.

Rose, Nikolas. 2009. "Normality and Pathology in a Biomedical Age." *Sociological Review* 57, no. 2: 66–83.

– 2013. "The Human Sciences in a Biological Age." *Theory, Culture and Society* 30, no. 1: 3–34.

– 2019. *Our Psychiatric Future*. Cambridge, UK: Polity Press.

Rose, Nikolas, and Joelle M. Abi-Rached. 2013. *Neuro: The New Brain Sciences and the Management of the Mind*. Princeton, NJ: Princeton University Press.

Rosenthal, Robert. 1979. "The File Drawer Problem and Tolerance for Null Results." *Psychological Bulletin* 86, no. 3: 638–41.

Saey, Tina Hesman. 2009. "You Are Who You Are by Default." *Science News* 176, no. 2: 16–20.

Sample, Ian. 2015. "Complex Living Brain Simulation Replicates Sensory Rat Behaviour." *Guardian* (London), 8 October. https://www.theguardian.com/science/2015/oct/08/complex-living-brain-simulation-replicates-sensory-rat-behaviour.

Samuel, Arthur L. 1960. "Programming Computers to Play Games." In *Advances in Computers*, vol. 1, ed. Franz L. Alt, 165–92. New York: Academic Press.

Sanders, Jet G., Hao-Ting Wang, Jonathan Schooler, and Jonathan Smallwood. 2017. "Can I Get Me out of My Head? Exploring Strategies for Controlling the Self-Referential Aspects of the Mind-Wandering State during Reading." *Quarterly Journal of Experimental Psychology* 70, no. 6: 1053–62.

Schinzel, Britta. 2006. "The Body in Medical Imaging between Reality and Construction." *Poiesis and Praxis: International Journal of Technology Assessment and Ethics of Science* 4, no. 3: 185–98.

Schneider, Leonid. 2017. "Human Brain Project: Bureaucratic Success Despite Scientific Failure." *For Better Science*, 22 February. https://forbetterscience.com/2017/02/22/human-brain-project-bureaucratic-success-despite-scientific-failure.

– 2018. "Interview with Human Brain Project Director Katrin Amunts." *For Better Science*, 14 September. https://forbetterscience.com/2018/09/14/interview-with-human-brain-project-director-katrin-amunts.

Schneider, Tanja, and Steve Woolgar. 2015a. "Neuromarketing in the Making: Enactment and Reflexive Entanglement in an Emerging Field." *BioSocieties* 10, no. 4: 400–21.

– 2015b. "Neuroscience beyond the Laboratory: Neuro Knowledges, Technologies and Markets." *BioSocieties* 10, no. 4: 389–99.

Schull, Natasha D., and Caitlin Zaloom. 2011. "The Shortsighted Brain: Neuroeconomics and the Governance of Choice in Time." *Social Studies of Science* 41, no. 4: 515–38.

– 2017. "Algorithms as Culture: Some Tactics for the Ethnography of Algorithmic Systems." *Big Data and Society* 4, no. 2: 1–12. https://doi.org/10.1177/2053951717738104.

Seli, Paul, Michael J. Kane, Jonathan Smallwood, Daniel L. Schacter, David Maillet, Jonathan W. Schooler, and Daniel Smilek. 2018. "Mind-Wandering as a Natural Kind: A Family-Resemblances View." *Trends in Cognitive Sciences* 22, no. 6: 479–90.

Serres, Michel. *The Parasite*. Trans. Lawrence R. Schehr. Baltimore, MD: Johns Hopkins University Press, 1982.

Shi, Lin, Peipeng Liang, Yishan Luo, Kai Liu, Vincent C.T. Mok, Winnie C.W. Chu, Defeng Wang, and Kucheng Li. 2017. "Using Large-Scale Statistical Chinese Brain Template (Chinese2020) in Popular

Neuroimage Analysis Toolkits." *Frontiers in Human Neuroscience* 11, art. 414: 1–6. https://doi.org/10.3389/fnhum.2017.00414.

Shine, James M., Patrick G. Bissett, Peter T. Bell, Oluwasanmi Koyejo, Joshua H. Balsters, Krzystof J. Gorgolwewski, Craig A. Moodle, and Russell A. Poldrack. 2016. "The Dynamics of Functional Brain Networks: Integrated Network Stats during Cognitive Task Performance." *Neuron* 92, no. 2: 544–54.

Shoham, Shy, Daniel H. O'Connor, and Ronen Segev. 2006. "How Silent Is the Brain: Is There a 'Dark Matter' Problem in Neuroscience?" *Journal of Comparative Psychology A* 192, no. 8: 151–63.

Shulman, Gordon L., Maurizio Corbetta, Randy L. Buckner, Julie A. Fiez, Francis M. Miezin, Marcus E. Raichle, and Steven E. Petersen. 1997. "Common Blood Flow Changes across Visual Tasks: I. Increases in Subcortical Structures and Cerebellum but Not in Nonvisual Cortex." *Journal of Cognitive Neuroscience* 9, no. 5: 624–47.

Silva, Gabriel A. 2011. "The Need for the Emergence of Mathematical Neuroscience: Beyond Computation and Simulation." *Frontiers in Computational Neuroscience* 5, art. 51: 1–3. https://www.frontiersin.org/articles/10.3389/fncom.2011.00051/full.

Simmons, Joseph P., Leif D. Nelson, and Uri Simonsohn. 2011. "False-Positive Psychology: Undisclosed Flexibility in Data Collection and Analysis Allows Presenting Anything as Significant." *Psychological Science* 22, no. 11: 1359–66.

– 2018. "False-Positive Citations." *Perspectives on Psychological Science* 13, no. 2: 255–9.

Simonsohn, Uri, Leif D. Nelson, and Joseph P. Simmons. 2014. "p-Curve and Effect Size: Correcting for Publication Bias Using Only Significant Results." *Perspectives on Psychological Science* 9, no. 6: 666–81.

Singh, Jennifer. 2015. *Multiple Autisms: Spectrums of Advocacy and Genomic Science*. Minneapolis: University of Minnesota Press.

Singh Chawla, Dalmeet. 2017. "Big Names in Statistics Want to Shake Up Much-Maligned P Value." *Nature* 548: 16–17. https://doi.org/10.1038/nature.2017.22375.

Sismondo, Sergio. 1999. "Models, Simulations, and Their Objects." *Science in Context* 12, no. 2: 247–60.

– 2011. "Simulation as a New Style of Research: Iteration, Integration, and Instability." In *From Science to Computational Sciences: Studies in the History of Computing and Its Influence on Today's Sciences*, ed. G. Gramelsberger, 151–63. Zürich: Diaphanes.

Smallwood, Jonathan, and Jessica Andrews-Hanna. 2013. "Not All Minds that Wander Are Lost: The Importance of a Balanced Perspective on the Mind-Wandering State." *Frontiers in Psychology* 4, art. 441: 1–6. https://doi.org/10.3389/fpsyg.2013.00441.

Smith, Stephen M., Karla L. Miller, Gholamreza Salimi-Khorshidi, Matthew Webster, Christian F. Beckmann, Thomas E. Nichols, Joseph D. Ramsey, and Mark W. Woolrich. 2011. "Network Modelling Methods for fMRI." *NeuroImage* 54, no. 2: 875–91.

Smoller, Joel, Blake Temple, and Zeke Vogler. 2017. "An Instability of the Standard Model of Cosmology Creates the Anomalous Acceleration without Dark Energy." *Proceedings of the Royal Society A: Mathematical, Physical and Engineering Sciences* 473, no. 2207: 1–20. https://doi.org/10.1098/rspa.2016.0887.

Soares, José M., Ricardo Magalhães, Pedro S. Moreira, Alexandre Sousa, Edward Ganz, Adriana Sampaio, Victor Alves, Paulo Margues, and Nuno Sousa. 2016. "A Hitchiker's Guide to Functional Magnetic Resonance Imaging." *Frontiers in Neuroscience* 10, art. 515: 1–35. https://doi.org/10.3389/fnins.2016.00515.

Spellman, Barbara. 2015. "A Short (Personal) Future History of Revolution 2.0." *Perspectives on Psychological Science* 10, no. 6: 886–99.

Sporns, Olaf. 2011. "The Human Connectome: A Complex Network." *Annals of the New York Academy of Science* 1224, no. 1: 109–25.

Spreng, Nathan R., Kathy D. Gerlach, Gary R. Turner, and Daniel L. Schacter. 2015. "Autobiographical Planning and the Brain: Activation and Its Modulation by Qualitative Features." *Journal of Cognitive Neuroscience* 27, no. 11: 2147–57.

Star, Susan Leigh, and Karen Ruhleder. 1996. "Steps Toward an Ecology of Infrastructure: Design and Access for Large Information Spaces." *Information Systems Research* 7, no. 1: 111–34. https://doi.org/10.1287/isre.7.1.111.

Starosielski, Nicole. 2015. *The Undersea Network*. Durham, NC: Duke University Press.

Stelzer, Johannes, Gabriele Lohmann, Karsten Mueller, Tilo Buschmann, and Robert Turner. 2014. "Deficient Approaches to Human Neuroimaging." *Frontiers in Human Neurosciences* 8, art. 462: 1–16. https://doi.org/10.3389/fnhum.2014.00462.

Stevens, Hallam. 2013. *Life out of Sequence: A Data-Driven History of Bioinformatics*. Chicago: University of Chicago Press.

Suchman, Lucy. 2007. *Human-Machine Reconfigurations: Plans and Situated Actions*. 2nd ed. Cambridge, UK: Cambridge University Press.

Sundberg, Mikaela. 2006. "Credulous Modellers and Suspicious Experimentalists? Comparison of Model Output and Data in Meteorological Simulation Modelling." *Science and Technology Studies* 19, no. 1: 52–68.

Szucs, Daniel, and John P. Ioannidis. 2017. "Empirical Assessment of Published Effect Sizes and Power in the Recent Cognitive Neuroscience and Psychology Literature." *PLOS Biology* 15, no. 3: 1–18. https://doi.org/10.1371/journal.pbio.2000797.

Taghia, Jalil, Weidong Cai, Srikanth Ryali, John Kochalka, Jonathan Nicholas, Tianwen Chen, and Vinod Menon. 2018. "Uncovering Hidden Brain State Dynamics That Regulate Performance and Decision-Making during Cognition." *Nature Communications* 9, art. 2505: 1–19. https://doi.org/10.1038/s41467-018-04723-6.

Talairach, Jean, and G. Szikla. 1967. *Atlas d'anatomie stéréotaxique du télencéphale*. Paris: Masson & Cie.

Thomas, Jeff. 2009. "Controversy and Consensus." In *Practising Science Communication in the Information Age: Theorising Professional Practices*, ed. Richard Holliman, Jeff Thomas, Sam Smidt, Eileen Scanlon, and Elizabeth Whitelegg, 131–48. Oxford: Oxford University Press.

Uttal, William R. 2001. *The New Phrenology: The Limits of Localizing Cognitive Processes in the Brain*. Cambridge, MA: MIT Press.

Vago, David R., and Fadel Zeidan. 2016. "The Brain on Silent: Mind Wandering, Mindful Awareness, and States of Mental Tranquility." *Annals of the New York Academy of Sciences* 1373, no. 1: 96–113.

van Beest, Ilja, and Kipling D. Williams. 2006. "When Inclusion Costs and Ostracism Pays, Ostracism Still Hurts." *Journal of Personality and Social Psychology* 91, no. 5: 918–28.

Van Calster, Laurens, Arnaud D'Argembeau, Eric Salmon, Frédéric Peters, and Steve Majerus. 2017. "Fluctuations of Attentional Networks and Default Mode Network during the Resting State Reflect Variations in Cognitive States: Evidence from a Novel Resting-State Experience Sampling Method." *Journal of Cognitive Neuroscience* 29, no. 1: 95–113.

van den Heuvel, Martijn P., and Olaf Sporns. 2013. "Network Hubs in the Human Brain." *Trends in Cognitive Science* 17, no. 12: 683–96.

Van Noorden, Richard. 2014. "'Boot Camps' Teach Scientists Computing Skills." *Nature*, 3 September. https://www.nature.com/news/boot-camps-teach-scientists-computing-skills-1.15799.

von Neumann, John. 1966. *Theory of Self-Reproducing Automata*. Urbana: University of Illinois Press.

Vidal, Fernando. 2009a. "Brainhood, Anthropological Figure of Modernity." *History of the Human Sciences* 22, no. 1: 5–36.

– 2009b. "The Cerebral Subject and the Challenge of Neurodiversity." *BioSocieties* 4, no. 4: 425–45.

– 2011. "Fictional Film and the Cerebral Subject." In *Neurocultures: Glimpses into an Expanding Universe*, ed. Francisco Ortega and Fernando Vidal, 329–44. Frankfurt am Main: Peter Lang.

Vul, Edward, Christine Harris, Piotr Winkielman, and Harold Pashler. 2008. "Voodoo Correlations in Social Neuroscience." Draft paper for *Perspectives on Psychological Science*. Pre-circulated online and now inaccessible.

– 2009a. "Puzzlingly High Correlations in fMRI Studies of Emotion, Personality, and Social Cognition." *Perspectives on Psychological Science* 4, no. 3: 274–90.

– 2009b. "Voodoo Correlations in Social Neuroscience: Rebuttal and Rejoinder." *Ed Vul*. https://www.edvul.com/voodoorebuttal.php.

Vul, Edward, and Hal Pashler. 2012. "Voodoo and Circularity Errors." *NeuroImage* 62, no. 2: 945–8.

Wallisch, Pascal, Michael E. Lusignan, Marc D. Benayoun, Tanya I. Baker, Adam S. Dickey, and Nicholas G. Hatsopoulos. 2014. MATLAB® *for Neuroscientists: An Introduction to Scientific Computing in* MATLAB®. 2nd ed. Amsterdam: Academic Press.

Walter, William Grey. 1934. "Thought and Brain: A Cambridge Experiment." *Spectator*, 5 October, 478–9.

Wasserstein, Ronald L., and Nicola A. Lazar. 2016. "The ASA's Statement on *p*-Values: Context, Process, and Purpose." *American Statistician* 70, no. 2: 129–33.

Weisberg, Deena S., Frank C. Keil, Joshua Goodstein, Elizabeth Rawson, and Jeremy R. Gray. 2008. "The Seductive Allure of Neuroscience Explanations." *Journal of Cognitive Neuroscience* 20, no. 3: 470–7.

Weisberg, Deena S., Jordan C.V. Taylor, and Emily J. Hopkins. 2015. "Deconstructing the Seductive Allure of Neuroscience Explanations." *Judgment and Decision Making* 10, no. 5: 429–41.

Wiener, Norbert. 1948. *Cybernetics: Or Communication and Control in the Animal and the Machine*. Cambridge, MA: MIT Press.

– 1957. "Rhythms in Physiology with Particular Reference to Electroencephalography." *Proceedings of the Rudolf Virchow Medical Society* 16: 109–24.

Wing, Jeannette, and Jay Gould. 1979. "Severe Impairments of Social Interaction and Associated Abnormalities in Children: Epidemiology

and Classification." *Journal of Autism and Developmental Disorders* 9,
  no. 1: 11–29.

Winsberg, Eric. 2010. *Science in the Age of Computer Simulation.*
  Chicago: University of Chicago Press.

Yarkoni, Tal. 2009. "Big Correlations in Little Studies: Inflated fMRI
  Correlations Reflect Low Statistical Power-Commentary on Vul et al.
  (2009)." *Perspectives on Psychological Science* 4, no. 3: 294–8.

Yerys, Benjamin E., Evan M. Gordon, Danielle N. Abrams, Theodore D.
  Satterthwaite, Rachel Weinblatt, Kathryn F. Jankowski, John Strang,
  Lauren Kenworthy, William D. Gaillard, and Chandan J. Vaidya. 2015.
  "Default Mode Network Segregation and Social Deficits in Autism
  Spectrum Disorder: Evidence from Non-Medicated Children DMN in
  Children with ASD." *NeuroImage: Clinical* 9: 223–32.

Yeung, Andy Wai Kan, Tazuko K. Goto, and Keung Leung. 2017. "The
  Changing Landscape of Neuroscience Research, 2006–2015: A
  Bibliometric Study." *Frontiers in Neuroscience* 11, art. 120: 1–10.
  https://doi.org/10.3389/fnins.2017.00120.

Yong, Ed. 2012. "Replication Studies: Bad Copy." *Nature* 485: 298–300.
  https://www.nature.com/news/replication-studies-bad-copy-1.10634.

– 2016. "Can Neuroscience Understand Donkey Kong, Let Alone a Brain."
  The Atlantic, 2 June. https://www.theatlantic.com/science/archive/
  2016/06/can-neuroscience-understand-donkey-kong-let-alone-a-brain/
  485177.

Zhang, Dongyang, and Marcus E. Raichle. 2010. "Disease and the Brain's
  Dark Energy." *Nature Reviews Neurology* 6, no. 1: 15–28.

Index

Abend, Gabriel, 36 Abi-Rached, Joelle, 14, 152n2

activation: activation-based imaging, 76, 84, 88 (*see also* BOLD paradigm); activation-focused cerebral geography, 84; all-or-nothing principle of, 83; of brain structures, 8, 14, 126, 140n2; data, 83; in the brains of volunteers, 55; of the circuits of love, 36; co-activations of distinct neuronal populations, 67, 108; contrasting activations, 64–70, 84; correlations between behavioural measures and brain activations, 44; data of brain activation, 20; emotional states and brain activations, 56; highly selective, 66; patterns of, 126; in response to external stimuli, 78, 80; in the salmon brain, 49; significant, 31, 66, 83; sparsely distributed, 84; triggering, 75; unique areas of, 76

Adrian, Edgar Douglas, 77–8, 86

adult neurogenesis, 11. *See also* plasticity

affect, 19–20, 133, 141n8

AI brain scans, 123–7

AI winters, 3

algorithms: algorithmic assemblages, 26, 107, 122; cognitive, 25, 135; generative, 125; neural network, 127; as quasi-biological organisms, 127. *See also* neural network(s); techniques: algorithmic AlphaGo, 6–7, 129–31

always-on society, 97

American Statistical Association, 146n10

Amoore, Louise, 107

Amunts, Katrin, 100, 149n5

amygdala, 11

anatomical maps, 20

Andreasen, Nancy, 86

Andrews-Hanna, Jessica, 87

anthropology of the machine, 136

artificial intelligence (AI): agent-based, 133; brain scans, 123–7; cognitive neuroscience and, 12, 123, 124; cognitive states that inspire, 136; division between neuroscience and, 4; human